Springer Series in
SOLID-STATE SCIENCES 174

Springer Series in
SOLID-STATE SCIENCES

Series Editors:
M. Cardona   P. Fulde   K. von Klitzing   R. Merlin   H.-J. Queisser   H. Störmer

The Springer Series in Solid-State Sciences consists of fundamental scientific books prepared by leading researchers in the field. They strive to communicate, in a systematic and comprehensive way, the basic principles as well as new developments in theoretical and experimental solid-state physics.

Please view available titles in *Springer Series in Solid-State Sciences*
on series homepage http://www.springer.com/series/682

Shun-Qing Shen

Topological Insulators

Dirac Equation in Condensed Matters

With 54 Figures

 Springer

Prof. Dr. Shun-Qing Shen
Department of Physics
The University of Hong Kong
Pokfulam Road, Hong Kong

ISSN 0171-1873
ISBN 978-3-642-32857-2   ISBN 978-3-642-32858-9 (eBook)
DOI 10.1007/978-3-642-32858-9
Springer Heidelberg New York Dordrecht London

Library of Congress Control Number: 2012953278

Printed on acid-free paper

Springer is part of Springer Science+Business Media (www.springer.com)

Preface

In recent years, we have seen rapid emergence of topological insulators and superconductors. The field is an important advance of the well-developed band theory in solids since its birth in the 1920s. The band theory or Fermi liquid theory and Landau's theory of spontaneously broken symmetry are two themes for most collective phenomena in many-body systems, such as semiconductors and superconductors. Discovery of the integer and fractional quantum Hall effects in the 1980s opens a new window to explore the mystery of condensed matters: topological order has to be introduced to characterize a large class of quantum phenomena. Topological insulator is a triumph of topological order in condensed matter physics.

The book grew out of a series of lectures I delivered in an international school on "Topology in Quantum Matter" at Bangalore, India, in July 2011. The aim of this book is to provide an introduction of a large family of topological insulators and superconductors based on the solutions of the Dirac equation. I believe that the Dirac equation is a key to the door of topological insulators. It is a line that could thread all relevant topological phases from one to three dimensions and from insulators to superconductors or superfluids. This idea actually defines the scope of this book on topological insulators. For this reason, a lot of topics in topological insulators are actually not covered in this book, for example, the interacting systems and topological field theory. Also I have no ambition to review rapid developments of the whole field and consequently no intention to introduce all topics in this introductory book.

I would like to express my gratitude to my current and former group members, and various parts of the manuscript benefited from the contributions of Rui-Lin Chu, Huai-Ming Guo, Jian Li, Hai-Zhou Lu, Jie Lu, Hai-Feng Lv, Wen-Yu Shan, Rui Yu, Yan-Yang Zhang, An Zhao, Yuan-Yuan Zhao, and Bin Zhou. Especially I would like to thank Hai-Zhou Lu for critically reading the manuscript and replotting all figures. I benefited from numerous discussions and collaborations with Qian Niu, Jainendra K. Jain, Jun-Ren Shi, Zhong Fang, and Xin Wan on the relevant topics. I am grateful for the support and suggestions from Lu Yu while writing this book. Some of the results in this book were obtained in my research projects funded by Research Grant Council of Hong Kong.

Hong Kong, China   Shun-Qing Shen
June 2012

Contents

List of Abbreviations

ARPES: angle-resolved photoemission spectroscopy
BCS: Bardeen-Cooper-Schrieffer
BdG: Bogoliubov-de Gennes
DOS: density of states
ESP: equal spin pairing
FQHE: fractional quantum Hall effect
HH: heavy hole
IQHE: integer quantum Hall effect
LH: light hole
NMR: nuclear magnetic resonance
PHS: particle-hole symmetry
QAHE: quantum anomalous Hall effect
QSHE: quantum spin Hall effect
STM: scanning tunneling microscopy
TAI: topological Anderson insulator
TKNN: Thouless-Kohmoto-Nightingale-Nijs
TRS: time reversal symmetry

Note: elementary charge: $-e$ (e is positive)

Chapter 1
Introduction

Abstract Discovery of topological insulators is a triumph of topological orders in quantum matters. The confirmed topological phases include superfluid liquid ^{3}He, integer and fractional quantum Hall effect, and topological insulators.

Keywords Hall effect • Quantum Hall effect • Quantum spin Hall effect • Topological insulator • Superfluids • Dirac equation

1.1 From the Hall Effect to Quantum Spin Hall Effect

In 1879, the American physicist Edwin H. Hall observed an effect that now bears his name: he measured a voltage that arises from the deflected motion of charged particles in solids under external electric field and magnetic field [1]. Consider a two-dimensional sample subjected to a perpendicular magnetic field **B**. When charged particles go through the sample, the particles deflect their motion due to the Lorentz force and accumulate near the boundary. As a result the charge accumulation at the boundary produces an electric field **E**. In equilibrium, the Lorentz force on a moving charged particle becomes zero:

$$\mathbf{F} = q(\mathbf{E} + \mathbf{v} \times \mathbf{B}) = 0, \tag{1.1}$$

where **v** is the velocity of the particle and q the charge of the particle. The voltage difference between the two boundaries is $V_H = EW$ (W the width of the sample), and the electric current through the sample is $I = q\rho_e vW$ (ρ_e is the density of the charge carriers). The ratio of the voltage to the electric current is known as the Hall resistance:

$$R_H = \frac{V_H}{I} = \frac{B}{q\rho_e} \tag{1.2}$$

S.-Q. Shen, *Topological Insulators: Dirac Equation in Condensed Matters*,
Springer Series in Solid-State Sciences 174, DOI 10.1007/978-3-642-32858-9_1,
© Springer-Verlag Berlin Heidelberg 2012

which is linear in the magnetic field B. In practice, the Hall effect is used to measure the sign of charge carriers q, that is, the particle-like or hole-like charge carrier, and the density of charge carriers ρ_e in solids. It can be also used to measure the magnetic field.

In the following year after his discovery, Hall measured the Hall resistance in ferromagnetic or even in paramagnetic metal in a magnetic field and observed that the Hall resistance could have additional contribution other than the linear term in magnetic field [2]. It could be proportional to the magnetization M in a ferromagnetic metal, that is, the Hall effect persists even in the absence of a magnetic field. An empirical relation is given by

$$R_H = R_O B + R_A M, \tag{1.3}$$

which has been applied to many materials over a broad range of external field. The second term represents the contribution from the magnetization M. This part of resistance cannot be simply understood as a result of the Lorentz force on a charged particle. It has taken almost one century to explore its physical origin. The main reason seems to be that this effect involves the topology of the band structure in solids, which had been just formulated in the 1980s. In 1954, Karplus and Luttinger [3] proposed a microscopic theory and found that electrons acquire an additional group velocity when an external electric field is applied to a solid. The anomalous velocity is perpendicular to the electric field and could make contribution to the Hall conductance. Now the anomalous velocity is formulated to be related to the change of phase of the Bloch wave function, when an electric field drives them to evolve in the crystal momentum space, and to be dependent on the crystal Hamiltonian [4,5]. ✳ Generally speaking, the anomalous Hall effect can have either an extrinsic origin due to disorder-related spin-dependent scattering of the charge carriers, or an intrinsic origin due to spin-dependent band structure of conduction electrons, which can be expressed in terms of the Berry phase in the momentum space [6]. This effect originates from the coupling of electron's orbital motion to its spin, which is a relativistic quantum mechanical effect. A spin-orbit force or spin transverse force can be used to understand the spin-dependent scattering by either the impurities or band structure. When an electron moves in an external electric field, the electron experiences a transverse force, which is proportional to spin current of electron, instead of charge current as in the Lorentz force [7]. As a result, electrons with spin-up will deflect to one direction, while electrons with spin-down to the opposite direction. In a ferromagnetic metal, the magnetization will cause an imbalance in the population between the electrons with spin-up and spin-down and consequently lead to the anomalous Hall effect.

While the anomalous Hall resistance vanishes in the absence of an external magnetic field and magnetization in a paramagnetic metal, the spin-dependent deflected motion of electrons in solids can still lead to an observable effect, that is, the spin Hall effect. The spin version of the Hall effect was first proposed by the Russian physicists Dyakonov and Perel in 1971 [8,9]. It consists of spin accumulation on the lateral surfaces of a current-carrying sample, the signs of the spin orientations being opposite on two opposite boundaries. When the current direction is reversed, the direction of spin orientation is also reversed. At the beginning, theorists predicted that the spin accumulation is caused by the asymmetric scattering of electrons with spin-up and spin-down in impurity potentials, which is named as the extrinsic spin Hall effect [10]. In 2003, two independent groups demonstrated that the spin-orbit coupling in the band structure of electrons can produce the transverse spin current even without impurity scattering, which is dubbed as the intrinsic spin Hall effect [11, 12]. In the quantum Hall regime, the competition between the Zeeman splitting and spin-orbit coupling leads to the resonant spin Hall effect, in which a small current may induce a finite spin current and spin polarization [13]. The spin Hall effect has been observed experimentally in GaAs and InGaAs thin film [14] and spin light-emitted diode of p-n junction [15].

The discovery of the integer quantum Hall effect opens a new phase in the study of the Hall effects. In 1980, von Klitzing, Dorda, and Pepper discovered experimentally that in two-dimensional electron gas at semiconductor hetero-junction subjected to a strong magnetic field, the longitudinal conductance becomes zero while the quantum plateau of the Hall conductance appears at $\nu e^2 / h$ [16]. The prefactor is an integer ($\nu = 1, 2, \ldots$), known as the filling factor. The quantum Hall effect is a quantum mechanical version of the Hall effect in two dimensions. This effect is very well understood now and can be simply explained in terms of single-particle orbitals of an electron in a magnetic field [17]. It is known that the motion of a charged particle in a uniform magnetic field is equivalent to that of a simple harmonic oscillator in quantum mechanics, in which the energy levels are quantized to be $\left(n + \frac{1}{2}\right) \hbar \omega_c$ and $\omega_c = eB/m$ is the cyclotron frequency. The energy levels are called the Landau levels and are highly degenerate. When one Landau level is fully filled, the filling factor is $\nu = 1$, and the corresponding Hall conductance is e^2 / h. Now it is realized that the integer ν is actually a topological invariant and is insensitive to the geometry of system and interaction of electrons [18].

To understand further, physicists like to use a semiclassical picture to explain the quantization of the Hall conductance. For a charged particle in a uniform magnetic field, the particle cycles around the magnetic flux rapidly because of the Lorentz force. The radius of the cycle is given by the magnetic field $R_n = \sqrt{\frac{\hbar}{eB}(2n + 1)}$. When the particle is close to the boundary, the particle bounces back from the rigid boundary and skips along the boundary forward. As a result, it forms a conducting channel along the boundary, which is called the edge state. The group velocity of the particle in the bulk is much slower than the cyclotron velocity, and then the particles in the bulk are intended to be pinned or localized by impurities or disorders. However, the rapid-moving particles along the edge channel are not affected by the impurities or disorders and form a perfect one-dimensional conducting channel with a quantum conductance e^2 / h. Consider the Landau levels are discrete. Each Landau level will generate one edge channel. Consequently, the number of the filled Landau levels or the filling factor determines the quantized Hall conductance. Thus, the key feature of the quantum Hall effect is that all electrons in the bulk are localized and the electrons near the edges form a series of edge-conducting channels [19], which is a characteristic of a topological phase.

In 1982, Tsui, Stormer, and Gossard observed that in a sample with higher mobility, the quantum plateau appears at the filling factor ν as a rational fraction ($\nu = \frac{1}{3}, \frac{2}{3}, \frac{1}{5}, \frac{2}{5}, \frac{3}{5}, \frac{12}{5}, \cdots$) known as the fractional quantum Hall effect [20]. The fractional quantum Hall effect relies fundamentally on electron–electron interactions as well as the Landau quantization. Laughlin proposed that the $\nu = 1/3$ state is a new type of many-body condensate, which can be described by the Laughlin wave function [21]. The quasiparticles in the condensate carry fractional charge $e/3$ because of strong Coulomb interaction. The observed Hall conductance plateau is due to the localization of fractionally charged quasiparticles in the condensate, and the fractional quantum Hall effect can be regarded as the integer quantum Hall effect of these quasiparticles. In 1988, Jainendra K. Jain proposed that the quasiparticles can be regarded as a combination of electron charge and quantum magnetic flux, that is, composite fermions [22]. This picture is applicable to all the quantum plateaus observed in the fractional quantum Hall effect. Now it is well accepted that the fractional quantum Hall effect is a topological quantum phase of composite fermions, which breaks time reversal symmetry.

In 1988 Haldane proposed that the integer quantum Hall effect could be realized in a lattice system of spinless electrons in a periodic magnetic flux [23]. Though the total magnetic flux is zero, electrons are driven to form a conducting edge channel by the periodic magnetic flux. As there is no pure magnetic field, the quantum Hall conductance originates from the band structure of electrons in the lattice instead of the discrete Landau levels for those in a strong magnetic field. This is a version of the quantized anomalous Hall effect in the absence of an external field or Landau levels. Furthermore it was found that the role of periodic magnetic flux can be replaced by the spin-orbit coupling. The quantized anomalous Hall effect can be realized in a ferromagnetic insulator with strong spin-orbit coupling. The anomalous Hall effect persists in an insulating regime. The anomalous Hall conductance can be expressed in terms of the integral of the Berry curvature over the momentum space or the Chern number for fully filled bands [24]. The Haldane model makes it possible to have nonzero Chern number for an electron band without a magnetic field. Though there have been extensive investigations on this topic [25–27], this effect has not yet been observed.

The quantum spin Hall effect is a quantum version of the spin Hall effect or a spin version of the quantum Hall effect and can be regarded as a combination of two quantum anomalous Hall effects of spin-up and spin-down electrons with opposite chirality. Overall it has no charge Hall conductance, but a nonzero spin Hall conductance. In 2005, Kane and Mele generalized the Haldane model to a graphene lattice of spin-$\frac{1}{2}$ electrons with the spin-orbit coupling [28]. The strong spin-orbit coupling is introduced to replace the periodic magnetic flux in the Haldane model. This interaction looks like a spin-dependent magnetic field to electron spins. Different electron spins experience opposite spin-orbit force, that is, spin transverse force [7]. As a result, a bilayer spin-dependent Haldane model may be realized in a spin-$\frac{1}{2}$ electron system with spin-orbit coupling, which exhibits the quantum spin Hall effect. In the case there exist spin-dependent edge states around the boundary of the system: electrons with different spins move in opposite directions and form a pair

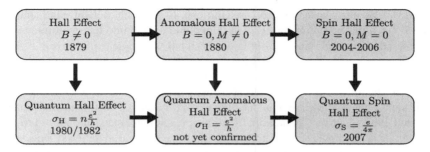

Fig. 1.1 Evolution from the ordinary Hall effect to the quantum spin Hall effect or two-dimensional topological insulator. Here, B stands for a magnetic field, and M stands for magnetization in a ferromagnet. The year means that the effect was discovered experimentally. σ_H is the Hall conductance, and σ_S is the spin Hall conductance

of helical edge states. Time reversal symmetry is still preserved, and the edge states are robust against impurities or disorders because the electron backscattering in the two edge channels is prohibited due to the symmetry. However, it was found that the spin-orbit coupling is very tiny in graphene. In 2006, Bernevig, Hughes, and Zhang proposed that the quantum spin Hall effect can be realized in the CdTe/HgTe/CdTe sandwiched quantum well [29]. HgTe is a material with an inverted band structure, and CdTe has a normal band structure. Tuning the thickness of HgTe layer may lead to the band inversion in the quantum well, which exhibits a topological phase transition. This prediction was confirmed experimentally by König et al. in the following year of the prediction [30]. The stability of quantum spin Hall effect was studied by several groups [31–34]. Li et al. discovered that the disorder may even generate the quantum spin Hall effect, and proposed a possible realization of topological Anderson insulator, in which all bulk electrons are localized by impurities, meanwhile a pair of conducting helical edge channels appear [35]. This phase was studied numerically and analytically [36,37]. Strong Coulomb interaction may also generate the quantum spin Hall effect in Mott insulators [38,39].

The quantum spin Hall effect is also dubbed as two-dimensional topological insulator. A flow chart from the ordinary Hall effect to the quantum spin Hall effect or two-dimensional topological insulator is presented in Fig. 1.1.

1.2 Topological Insulators as Generalization of Quantum Spin Hall Effect

There is no Hall effect in three dimensions. However, the generalization of the quantum spin Hall effect to three dimensions is one of the milestones in the development of topological insulators [40–43]. It is not a simple generalization of transverse transport of electron charge or spin from two dimensions to three dimensions, or the Hall effect. Instead it is the evolution of the bound states near the system boundary based on the intrinsic band structure: the one-dimensional helical edge states in two-dimensional quantum spin Hall system could evolve into the two-dimensional surface states surrounding the three-dimensional topological insulators. A topological insulator is a state of quantum matter that behaves as an insulator in its interior while as a metal on its boundary. In the bulk of a topological insulator, the electronic band structure resembles an ordinary insulator, with separated conduction and valence bands. Near the boundary, the surface states exist within the bulk energy gap and allow electron conduction. Electron spins in these states are locked to their momenta. A topological insulator preserves time reversal symmetry. Due to the Kramers degeneracy, at a given energy there always exists a pair of states that have opposite spins and momenta, so the backscattering between these states is strongly suppressed. These states are characterized by a topological index. Kane and Mele proposed a Z_2 index to classify the materials with time reversal invariance into strong and weak topological insulators [44]. For a weak topological insulator, the resultant surface states are not so stable to disorder or impurities, although its physics is very similar to that in two-dimensional states. A strong topological insulator has more subtle relation to the quantum spin Hall system. It is possible to classify the conventional insulator and topological insulator by time reversal symmetry. The surface states in a strong topological insulator are protected by time reversal symmetry.

$Bi_{1-x}Sb_x$ was the first candidate for three-dimensional topological insulator predicted by Fu and Kane [45] and verified experimentally soon after the prediction [46]. Zhang et al. [47] and Xia et al. [48] pointed out that Bi_2Te_3 and Bi_2Se_3 are topological insulators with a single Dirac cone of the surface states. Angle-resolved photoemission spectroscopy (ARPES) data showed clearly the existence of a single Dirac cone in Bi_2Se_3 [48] and Bi_2Te_3 [49]. Electrons in the surface states possess a quantum spin texture, and electron momenta are coupled strongly with electron spins. These may result in a lot of exotic magnetoelectric properties. Qi et al. [50] proposed the unconventional magnetoelectric effect for the surface states, in which electric and magnetic fields are coupled together and are governed by the so-called "axion equation" instead of Maxwell's equations. It is regarded as one of the characteristic features of the topological insulators [51, 52]. Fu and Kane proposed possible realization of Majorana fermions as a proximity effect of s-wave superconductor and surface states of topological insulator [53]. Majorana fermions are topologically protected from local sources of decoherence and will be of potential application in universal quantum computation [54, 55]. Thus, the topological insulators provide a new platform to explore novel and exotic quantum particles in condensed matters.

Reduction of dimensionality to one dimension brings some new insights in one-dimensional systems with topological properties. The boundary of one-dimensional system is simply an end point. A one-dimensional topological insulator is an insulator with two end states of zero energy. Study of the end states in one dimension has dated back to the 1980s. The polyacetylene is a one-dimensional organic material with the so-called A and B phases. It was realized that the domain walls connecting the A and B phase induce rigid solitons with zero energy and are the charge carriers for this organic conductor [56]. While the soliton and antisoliton are topological excitations in polyacetylene, the A and B phases are actually topologically distinct for an open boundary condition: one phase possesses two end states of zero energy, while the other does not, although both phases open an energy gap due to the Peierls instability or dimerization of the lattice. We shall demonstrate that this is actually the simplest topological insulator in one dimension.

1.3 Topological Phases in Superconductors and Superfluids

Liquid helium ^{3}He has two different superfluid phases at low temperatures, the A and B phases. The ^{3}He atoms are fermions of charge neutral and can be described by the conventional Fermi liquid theory just like electrons in metal. Osheroff, Lee, and Richardson [57] studied the pressurization curve of a mixture of liquid and solid ^{3}He and observed two reproducible anomalies, which indicate that the liquid phase existing between 2.0 and 2.6 mK is the A phase and that below 2.0 mK is the B phase. The normal to A phase transition at $T_A \sim 2.6$ mK is of second order and A-B phase transition at $T_B \sim 2$ mK is of first order. The theory of superconductivity for electrons in spin-triplet states was first developed by Balian and Werthamer in 1963 [58]. They observed that all Cooper pairs are in the p-wave pairing ($l = 1$) and spin-triplet states, which succeeds in explaining superfluidity in the B phase. The pairing symmetry determines the topology of the band structure of quasiparticles. The A phase has different topology from the B phase: the pairs form only in the state of $S_z = 1$ and/or $S_z = -1$, that is, the so-called equal-spin pairing state or Anderson-Brinkman-Morel state. This conclusion was first drawn from an analysis of the spatial profile of nuclear magnetic resonance (NMR) experiment [59].

Physics we learned from the superfluid phase of liquid ^{3}He has been widely applied in various fields from particle physics and cosmology to condensed matter physics [60]. In the spinless p-wave pairing superconductor, there exist weak and strong pairing phases, which are characterized by different topological invariants. The weak pairing phase is topologically nontrivial and may have chiral edge states around the boundary of system, very similar to that in the quantum Hall effect [61]. After the discovery of the fractional quantum Hall effect, it is found that the weak pairing state has a pairing wave function which is asymptotically the same as in the Moore-Read quantum Hall state. Thus, the topological order was introduced to characterize the superfluid phases. The topological aspects in these two phases have been discussed in details in the book by Volovik [60]. Some concepts and topological invariants can be applied explicitly to topological insulators in the framework of a single-particle wave functions in the band theory. For example, the band inversion could accompany a topological quantum phase transition as in the quantum spin Hall effect.

Now we have realized that the Bogoliubov-de Gennes equation for superconductors and superfluids has very similar/identical mathematical structure as the Dirac equation for topological insulators. Like the band gap in insulators, quasiparticles in superconductors and superfluids may also have a nonzero gap. The symmetry classification of noninteracting Hamiltonian emerged in the context of random matrix theory long before the discovery of topological insulators. Schnyder et al. [62] systematically studied the topological phases in insulators and superconductors and provided an exhaustive classification of topological insulators and superconductors for noninteracting systems of fermions. Bogoliubov-de Gennes equation has particle-hole symmetry, and the Dirac equation has time reversal symmetry. Similarity between particle-hole symmetry and time reversal symmetry makes it possible to study the topological insulators and superconductors in a single framework.

Discovery of topological insulator stimulates to reexamine the properties of spin-triplet superconductors which are potential candidates for topological superconductors. Among several classes of spin-triplet superconductors, Sr_2RuO_4 is thought to be a p-wave pairing superconductor and similar to the A phase in superfluid liquid ^3He [63, 64]. Initial data in the measurement of tunneling spectroscopy suggest possible existence of chiral edge states in Sr_2RuO_4 [65]. Cu-doped topological insulator $Cu_xBi_2Se_3$, which becomes superconducting below $T_c = 3.8$ K [66], is also suggested to be a topological superconductor [67].

1.4 Dirac Equation and Topological Insulators

The Dirac equation is a relativistic quantum mechanical one for elementary spin-$\frac{1}{2}$ particle [68,69]. It enters the field of topological insulator in two aspects. First of all, a large class of topological insulators possesses strong spin-orbit coupling, which is a consequence of the Dirac equation in nonrelativistic limit [70]. It makes the spin, the momentum, and the Coulomb interaction or external electric fields couple together. As a result, it is possible that the band structures in some materials become topologically nontrivial. This provides a physical origin to form a topological insulator. The other aspect is that the effective Hamiltonians for the quantum spin Hall effect and three-dimensional topological insulators have the identical mathematical structure of the Dirac equation. In these effective models the equation is employed to describe the coupling between electrons of the conduction and valence bands in semiconductors, not the electrons and positrons in Dirac theory. The positive and negative spectra are for the electrons and holes in semiconductors, respectively, not those in the high-energy physics. The conventional Dirac equation is time reversal invariant. For a system with time reversal symmetry, the effective Hamiltonian to describe the electrons near the Fermi level can be derived from the theory of invariants or the $k \cdot p$ theory. As a consequence of the $k \cdot p$ expansion of the band structure, some effective continuous models have the identical mathematical structure as the Dirac equation. The equation can be also obtained from the effective model near the critical point of topological quantum phase transition.

Generally speaking, each topological insulator or superconductor is governed by one Dirac equation. In this book we get started with the Dirac equation to provide a simple but unified description for a large family of topological insulators and superconductors. A series of solvable differential equations are presented to demonstrate the existence of the end, edge, and surface states in topological matters.

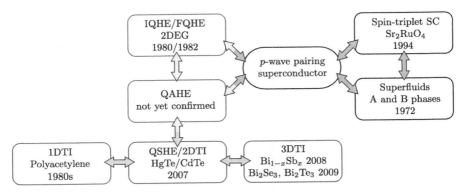

Fig. 1.2 The family of topological insulators and superconductors. TI stands for topological insulator, SC for superconductor, IQHE/FQHE for the integer and fractional quantum Hall effect, QAHE for quantum anomalous Hall effect, and QSHE for quantum spin Hall effect. The materials are followed by the year of discovery. Edge states in Sr_2RuO_4 need to be confirmed further

1.5 Summary: The Confirmed Family Members

As a summary, we list the confirmed topological insulators and superconductors in Fig. 1.2. There are three streams in the development of the field:

1. The Hall effect: the integer and fractional quantum Hall effects (1980, 1982), the quantum anomalous Hall effect (predicted in 1988, not yet confirmed experimentally), and the quantum spin Hall effect (2007)
2. Topological insulators: one-dimensional polyacetylene (1980s); two-dimensional HgTe/CdTe quantum well (2007), InAs/GaSb quantum well (2011); and three-dimensional $Bi_{1-x}Sb_x$ (2008), Bi_2Te_3 (2009), Bi_2Se_3 (2009), Bi_2Te_2Se (2010),. . ..
3. The p-wave superconductor: superfluid A and B phases in liquid ^3He (1972) and equal-spin pairing (ESP) superconductor such as Sr_2RuO_4 (highly possible)

1.6 Further Reading

Introductory Materials:

- J.E. Moore, The birth of topological insulators. Nature (London) **464**, 194 (2010)
- X.L. Qi, S.C. Zhang, The quantum spin Hall effect and topological insulators. Phys. Today **63**, 33 (2010)

Overview:

- M.Z. Hasan, C.L. Kane, Topological insulators. Rev. Mod. Phys. **82**, 3045 (2010)
- X.L. Qi, S.C. Zhang, Topological insulators and superconductors. Rev. Mod. Phys. **83**, 1057 (2011)

References

1. E.H. Hall, Am. J. Math. **2**, 287 (1879)
2. E.H. Hall, Philos. Mag. **12**, 157 (1881)
3. R. Karplus, J.M. Luttinger, Phys. Rev. **95**, 1154 (1954)
4. M.C. Chang, Q. Niu, Phys. Rev. Lett. **75**, 1348 (1995)
5. D. Xiao, M.C. Chang, Q. Niu, Rev. Mod. Phys. **82**, 1959 (2010)
6. N. Nagaosa, J. Sinova, S. Onoda, A.H. MacDonald, N.P. Ong, Rev. Mod. Phys. **82**, 1539 (2010)
7. S.Q. Shen, Phys. Rev. Lett. **95**, 187203 (2005)
8. M.I. Dyakonov, V.I. Perel, JETP Lett. **13**, 467 (1971)
9. M.I. Dyakonov, V.I. Perel, Phys. Lett. A **35**, 459 (1971)
10. J.E. Hirsch, Phys. Rev. Lett. **83**, 1834 (1999)
11. S. Murakami, N. Nagaosa, S.C. Zhang, Science **301**, 1348 (2003)
12. J. Sinova, D. Culcer, Q. Niu, N.A. Sinitsyn, T. Jungwirth, A.H. MacDonald, Phys. Rev. Lett. **92**, 126603 (2004)
13. S.Q. Shen, M. Ma, X.C. Xie, F.C. Zhang, Phys. Rev. Lett. **92**, 256603 (2004)
14. Y.K. Kato, R.C. Myers, A.C. Gossard, D.D. Awschalom, Science **306**, 1910 (2004)
15. J. Wunderlich, B. Kaestner, J. Sinova, T. Jungwirth, Phys. Rev. Lett. **94**, 047204 (2005)
16. K. von Klitzing, G. Dorda, M. Pepper, Phys. Rev. Lett. **45**, 494 (1980)
17. R.B. Laughlin, Phys. Rev. B **23**, 5632 (1981)
18. D.J. Thouless, M. Kohmoto, M.P. Nightingale, M. den Nijs, Phys. Rev. Lett. **49**, 405 (1982)
19. B.I. Halperin, Phys. Rev. B **25**, 2185 (1982)
20. D.C. Tsui, H.L. Stormer, A.C. Gossard, Phys. Rev. Lett. **48**, 1559 (1982)
21. R.B. Laughlin, Phys. Rev. Lett. **50**, 1395 (1983)
22. J.K. Jain, Phys. Rev. Lett. **63**, 199 (1989)
23. F.D.M. Haldane, Phys. Rev. Lett. **61**, 2015 (1988)
24. T. Jungwirth, Q. Niu, A.H. MacDonald, Phys. Rev. Lett. **88**, 207208 (2002)
25. M. Onoda, N. Nagaosa, Phys. Rev. Lett. **90**, 206601 (2003)
26. C.X. Liu, X.L. Qi, X. Dai, Z. Fang, S.C. Zhang, Phys. Rev. Lett. **101**, 146802 (2008)
27. R. Yu, W. Zhang, H. Zhang, S. Zhang, X. Dai, Z. Fang, Science **329**, 61 (2010)
28. C.L. Kane, E.J. Mele, Phys. Rev. Lett. **95**, 226801 (2005)
29. B.A. Bernevig, T.L. Hughes, S.C. Zhang, Science **314**, 1757 (2006)
30. M. König, S. Wiedmann, C. Brüne, A. Roth, H. Buhmann, L.W. Molenkamp, X.L. Qi, S.C. Zhang, Science **318**, 766 (2007)
31. D.N. Sheng, Z.Y. Weng, L. Sheng, F.D.M. Haldane, Phys. Rev. Lett. **97**, 036808 (2006)
32. C. Xu, J.E. Moore, Phys. Rev. B **73**, 045322 (2006)
33. C. Wu, B.A. Bernevig, S.C. Zhang, Phys. Rev. Lett. **96**, 106401 (2006)
34. M. Onoda, Y. Avishai, N. Nagaosa, Phys. Rev. Lett. **98**, 076802 (2007)
35. J. Li, R.L. Chu, J.K. Jain, S.Q. Shen, Phys. Rev. Lett. **102**, 136806 (2009)
36. H. Jiang, L. Wang, Q.F. Sun, X.C. Xie, Phys. Rev. B **80**, 165316 (2009)
37. C.W. Groth, M. Wimmer, A.R. Akhmerov, J. Tworzydło, C.W.J. Beenakker, Phys. Rev. Lett. **103**, 196805 (2009)
38. R.S.K. Mong, A.M. Essin, J.E. Moore, Phys. Rev. B **81**, 245209 (2010)
39. D. Pesin, L. Balents, Nat. Phys. **6**, 376 (2010)
40. L. Fu, C.L. Kane, E.J. Mele, Phys. Rev. Lett. **98**, 106803 (2007)
41. J.E. Moore, L. Balents, Phys. Rev. B **75**, 121306(R) (2007)
42. S. Murakami, New. J. Phys. **9**, 356 (2007)
43. R. Roy, Phys. Rev. B **79**, 195322 (2009)
44. C.L. Kane, E.J. Mele, Phys. Rev. Lett. **95**, 146802 (2005)
45. L. Fu, C.L. Kane, Phys. Rev. B **76**, 045302 (2007)
46. D. Hsieh, D. Qian, L. Wray, Y. Xia, Y.S. Hor, R.J. Cava, M.Z. Hasan, Nature (London) **452**, 970 (2008)
47. H. Zhang, C.X. Liu, X.L. Qi, X. Dai, Z. Fang, S.C. Zhang, Nat. Phys. **5**, 438 (2009)

48. Y. Xia, D. Qian, D. Hsieh, L. Wray, A. Pal, H. Lin, A. Bansil, D. Grauer, Y.S. Hor, R.J. Cava, M.Z. Hasan, Nat. Phys. **5**, 398 (2009)
49. Y.L. Chen, J.G. Analytis, J.H. Chu, Z.K. Liu, S.K. Mo, X.L. Qi, H.J. Zhang, D.H. Lu, X. Dai, Z. Fang, S.C. Zhang, I.R. Fisher, Z. Hussain, Z.X. Shen, Science **325**, 178 (2009)
50. X.L. Qi, T.L. Hughes, S.C. Zhang, Phys. Rev. B **78**, 195424 (2008)
51. X.L. Qi, R.D. Li, J.D. Zang, S.C. Zhang, Science **323**, 1184 (2009)
52. A.M. Essin, J.E. Moore, D. Vanderbilt, Phys. Rev. Lett. **102**, 146805 (2009)
53. L. Fu, C.L. Kane, Phys. Rev. Lett. **100**, 096407 (2008)
54. M.H. Freedman, M. Larsen, Z. Wang, Commun. Math. Phys. **227**, 605 (2002)
55. A. Kitaev, Ann. Phys. (N.Y.) **321**, 2 (2006)
56. A.J. Heeger, S. Kivelson, J.R. Schrieffer, W.P. Su, Rev. Mod. Phys. **60**, 781 (1988)
57. D.D. Osheroff, R.C. Richardson, D.M. Lee, Phys. Rev. Lett. **28**, 885 (1972)
58. R. Balian, N.R. Werthamer, Phys. Rev. **131**, 1553 (1963)
59. A.J. Leggett, Rev. Mod. Phys. **76**, 999 (2004)
60. G.E. Volovik, *The Universe in a Helium Droplet* (Clarendon, Oxford, 2003)
61. N. Read, D. Green, Phys. Rev. B **61**, 10267 (2000)
62. A.P. Schnyder, S. Ryu, A. Furusaki, A.W.W. Ludwig, Phys. Rev. B **78**, 195125 (2008)
63. K. Ishida, H. Mukuda, Y. Kitaoka, K. Asayama, Z.Q. Mao, Y. Mori, Y. Maeno, Nature (London) **396**, 658 (1998)
64. T.M. Rice, M. Sigrist, J. Phys. Condens. Matter **7**, L643 (1995)
65. S. Kashiwaya, H. Kashiwaya, H. Kambara, T. Furuta, H. Yaguchi, Y. Tanaka, Y. Maeno, Phys. Rev. Lett. **107**,077003 (2011)
66. Y.S. Hor, A.J. Williams, J.G. Checkelsky, P. Roushan, J. Seo, Q. Xu, H.W. Zandbergen, A. Yazdani, N.P. Ong, R.J. Cava, Phys. Rev. Lett. **104**, 057001 (2010)
67. L. Fu, E. Berg, Phys. Rev. Lett. **105**, 097001 (2010)
68. P.A.M. Dirac, Proc. R. Soc. A **117**, 610 (1928)
69. P.A.M. Dirac, *Principles of Quantum Mechanics*, 4th edn. (Clarendon, Oxford, 1982)
70. R. Winkler, *Spin-Orbit Coupling Effects in Two-Dimensional Electrons and Hole Systems* (Springer, Berlin, 2003)

Chapter 2
Starting from the Dirac Equation

Abstract The Dirac equation is a key to the door of topological insulators and superconductors. A quadratic correction to the equation makes it topologically distinct. The solution of the bound states near the boundary reflects the topology of the band structure of the system.

Keywords The modified Dirac equation • Solution of the bound states • End state • Edge state • Surface state

2.1 Dirac Equation

In 1928, Paul A.M. Dirac wrote down an equation for a relativistic quantum mechanical wave function, which describes an elementary spin-$\frac{1}{2}$ particle [1, 2]:

$$H = c\mathbf{p} \cdot \alpha + mc^2\beta \tag{2.1}$$

where m is the rest mass of particle and c is the speed of light. α_i and β are known as the Dirac matrices satisfying the relations

$$\alpha_i^2 = \beta^2 = 1, \tag{2.2a}$$

$$\alpha_i\alpha_j = -\alpha_j\alpha_i, \tag{2.2b}$$

$$\alpha_i\beta = -\beta\alpha_i. \tag{2.2c}$$

Here a_i and β are not simple complex numbers. The anticommutation relation means that they can obey a Clifford algebra and must be expressed in a matrix form. In one- and two-dimensional spatial space, they are at least 2×2 matrices. The Pauli matrices σ_i ($i = x, y, z$) satisfy all these relations:

$$\{\sigma_i, \sigma_j\} = 2\delta_{ij}, \tag{2.3}$$

S.-Q. Shen, *Topological Insulators: Dirac Equation in Condensed Matters*,
Springer Series in Solid-State Sciences 174, DOI 10.1007/978-3-642-32858-9_2,
© Springer-Verlag Berlin Heidelberg 2012

where

$$\sigma_x = \begin{pmatrix} 0 & 1 \\ 1 & 0 \end{pmatrix}, \sigma_y = \begin{pmatrix} 0 & -i \\ i & 0 \end{pmatrix}, \sigma_z = \begin{pmatrix} 1 & 0 \\ 0 & -1 \end{pmatrix}. \tag{2.4}$$

Thus, in one dimension, the two Dirac matrices α_x and β are any two of the three Pauli matrices, for example,

$$\alpha_x = \sigma_x, \beta = \sigma_z. \tag{2.5}$$

In two dimensions, the three Dirac matrices are the Pauli matrices:

$$\alpha_x = \sigma_x, \alpha_y = \sigma_y, \beta = \sigma_z. \tag{2.6}$$

In three dimensions, we cannot find more than three 2×2 matrices satisfying the anticommutation relations. Thus, the four Dirac matrices are at least four-dimensional and can be expressed in terms of the Pauli matrices

$$\alpha_i = \begin{pmatrix} 0 & \sigma_i \\ \sigma_i & 0 \end{pmatrix} \equiv \sigma_x \otimes \sigma_i, \tag{2.7a}$$

$$\beta = \begin{pmatrix} \sigma_0 & 0 \\ 0 & -\sigma_0 \end{pmatrix} \equiv \sigma_z \otimes \sigma_0 \tag{2.7b}$$

where σ_0 is a 2×2 identity matrix.

From this equation, the relativistic energy-momentum relation will be automatically the solution of the equation

$$E^2 = m^2 c^4 + p^2 c^2. \tag{2.8}$$

In three dimensions, one has two solutions for positive energy E_+ and two solutions for negative energy, E_-:

$$E_\pm = \pm \sqrt{m^2 c^4 + p^2 c^2}. \tag{2.9}$$

This equation can be used to describe the motion of an electron with spin: the two solutions of the positive energy correspond to two states of electron with spin-up and spin-down, while the two solutions of the negative energy correspond to a positron with spin-up and spin-down. The energy gap between these two particles is $2mc^2 (\approx 1.0 \, \text{MeV})$.

This equation demands the existence of antiparticle, that is, a particle with negative energy or mass, and predates the discovery of positron, the antiparticle of electron. It is one of the main achievements of modern theoretical physics. Dirac proposed that the negative energy states are fully filled, in which the Pauli exclusion principle prevents a particle transiting into such occupied states. The normal state of the vacuum then consists of an infinite density of negative energy states. The state for a single electron means that all the states of negative energies are filled and only one state of positive energy is filled. It is assumed that deviation from the norm produced by employing one or more of the negative energy states can be observed. The absence of a negative charged electron that has a negative mass and kinetic energy would be then expected to manifest itself as a positively charged particle which has an equal positive mass and positive energy. In this way, a "hole" or positron can be formulated. Unlike the Schrödinger equation for a single particle, the Dirac theory in principle is a many-body theory. This has been discussed in many textbooks on quantum mechanics [2].

Under the transformation of mass $m \rightarrow -m$, it is found that the equation remains invariant if we replace $\beta \rightarrow -\beta$, which satisfies all mutual anticommutation relations for α_i and β in Eq. (2.2). This reflects the symmetry between the positive and negative energy particles in the Dirac equation: there is no topological distinction between particles with positive and negative masses.

2.2   Solutions of Bound States

2.2.1   Jackiw-Rebbi Solution in One Dimension

Possible relation between the Dirac equation and the topological insulator reveals from a solution of the bound state at the interface between two regions of positive and negative masses. We get started with

$$h(x) = -iv\hbar\partial_x\sigma_x + m(x)v^2\sigma_z \tag{2.10}$$

and

$$m(x) = \begin{cases} -m_1 & \text{if} \quad x < 0 \\ +m_2 & \text{otherwise} \end{cases} \tag{2.11}$$

(and m_1 and $m_2 > 0$). We use an effective velocity v to replace the speed of light c when the Dirac equation is applied to solids. The eigenvalue equation has the form

$$\begin{pmatrix} m(x)v^2 & -iv\hbar\partial_x \\ -iv\hbar\partial_x & -m(x)v^2 \end{pmatrix} \begin{pmatrix} \varphi_1(x) \\ \varphi_2(x) \end{pmatrix} = E \begin{pmatrix} \varphi_1(x) \\ \varphi_2(x) \end{pmatrix}. \tag{2.12}$$

For either $x < 0$ or $x > 0$, the equation is a second-order ordinary differential equation. We can solve the equation at $x < 0$ and $x > 0$ separately. The solution of the wave function should be continuous at $x = 0$. In order to have a solution of a bound state near the junction, we take the Dirichlet boundary condition that the wave function must vanish at $x = \pm\infty$. For $x > 0$, we set the trial wave function

$$\begin{pmatrix} \varphi_1(x) \\ \varphi_2(x) \end{pmatrix} = \begin{pmatrix} \varphi_1^+ \\ \varphi_2^+ \end{pmatrix} e^{-\lambda_+ x}. \tag{2.13}$$

Then the secular equation gives

$$\det \begin{pmatrix} m_2v^2 - E & iv\hbar\lambda_+ \\ iv\hbar\lambda_+ & -m_2v^2 - E \end{pmatrix} = 0. \tag{2.14}$$

The solution to this equation is $\lambda_+ = \pm\sqrt{m_2^2v^4 - E^2}/v\hbar$.

The solutions λ can be either real or purely imaginary. For $m_2^2v^4 < E^2$, the solutions are purely imaginary, and the corresponding wave function spreads over the whole space. They are the extended states or the bulk states which we are not interested here. For $m_2^2v^4 > E^2$, the solutions are real, and we choose positive λ_+ to satisfy the boundary condition at $x \to +\infty$. The two components in the wave function satisfy

$$\varphi_1^+ = -\frac{iv\hbar\lambda_+}{m_2v^2 - E}\varphi_2^+. \tag{2.15}$$

Similarly, for $x < 0$, we have

$$\begin{pmatrix} \varphi_1(x) \\ \varphi_2(x) \end{pmatrix} = \begin{pmatrix} \varphi_1^- \\ \varphi_2^- \end{pmatrix} e^{+\lambda_- x} \tag{2.16}$$

with $\lambda_- = \sqrt{m_1^2v^4 - E^2}/v\hbar$, and

$$\varphi_1^- = -\frac{iv\hbar\lambda_-}{m_1v^2 + E}\varphi_2^-. \tag{2.17}$$

At $x = 0$, the continuity condition for the wave function requires

$$\begin{pmatrix} \varphi_1^+ \\ \varphi_2^+ \end{pmatrix} = \begin{pmatrix} \varphi_1^- \\ \varphi_2^- \end{pmatrix}. \tag{2.18}$$

From this equation, it follows that

$$\frac{-\sqrt{m_2^2v^4 - E^2}}{m_2v^2 - E} = \frac{\sqrt{m_1^2v^4 - E^2}}{-m_1v^2 - E}. \tag{2.19}$$

Thus, there exists a solution of zero energy $E = 0$, and the corresponding wave function is

$$\Psi(x) = \sqrt{\frac{v}{\hbar}\frac{m_1m_2}{m_1 + m_2}} \begin{pmatrix} 1 \\ i \end{pmatrix} e^{-|m(x)vx|/\hbar}. \tag{2.20}$$

The solution dominantly distributes near the interface or domain wall at $x = 0$ and decays exponentially away from the original point $x = 0$, as shown in Fig. 2.1. The solution of $m_1 = m_2$ was first obtained by Jackiw and Rebbi and is a mathematical basis for existence of topological excitations or solitons

Fig. 2.1 The probability density $|\Psi(x)|^2$ of the solution as a function of position in Eq. (2.20)

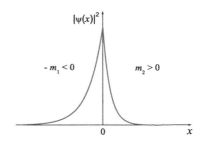

in one-dimensional systems [3]. The spatial distribution of the wave function is determined by the characteristic scales $\xi_{1,2} = \lambda_{\pm}^{-1} = \hbar/|m_{1,2}v|$. The solution exists even when $m_2 \to +\infty$. In this case, $\Psi(x) \to 0$ for $x > 0$. However, we have to point out that the wave function does not vanish at the interface, $x = 0$. If we regard the vacuum as a system with an infinite positive mass, a system of a negative mass with an open boundary condition possesses a bound state near the boundary. This result leads to some popular pictures for the formation of the edge and surface states in topological insulators.

Stability of the zero mode solution: we may find a general solution of zero energy for a distribution of mass $m(x)$ changing from negative to positive mass at two ends. Consider the solution of $E = 0$ to Eq. (2.10). The eigenvalue equation is reduced to

$$\left[-iv\hbar\partial_x\sigma_x + m(x)v^2\sigma_z\right]\varphi(x) = 0. \tag{2.21}$$

Multiplying σ_x from the left-hand side, we have

$$\partial_x\varphi(x) = -\frac{m(x)v}{\hbar}\sigma_y\varphi(x). \tag{2.22}$$

Thus, the wave function should be the eigenstate of σ_y,

$$\sigma_y\varphi_\eta(x) = \eta\varphi_\eta(x) \tag{2.23}$$

with

$$\varphi_\pm = \frac{1}{\sqrt{2}}\begin{pmatrix} 1 \\ \pm i \end{pmatrix}\varphi(x).$$

The wave function has the form

$$\varphi_\eta(x) \propto \frac{1}{\sqrt{2}}\begin{pmatrix} 1 \\ \eta i \end{pmatrix}\exp\left[-\int^x \eta\frac{m(x')v}{\hbar}dx'\right]. \tag{2.24}$$

For $x \to \pm\infty$, $\varphi(x) \propto \exp\left[-|m(\pm\infty)vx|/\hbar\right]$, in which the sign η is determined by the signs of $m(\pm\infty)$. If $m(+\infty)$ and $m(-\infty)$ differ by a sign as a domain wall, there always exists a zero-energy solution near a domain wall of the mass distribution $m(x)$. Therefore, this solution is quite robust against the mass distribution $m(x)$.

2.2.2 Two Dimensions

In two dimensions (with $p_z = 0$), we consider a system with an interface parallel
to the y-axis, with $m(x) = -m_1$ for $x < 0$ and m_2 for $x > 0$. $p_y = \hbar k_y$ is a
good quantum number. We have two solutions which the wave functions dominantly
distribute around the interface: one solution has the form

$$\Psi_+(x, k_y) = \sqrt{\frac{v}{h} \frac{m_1 m_2}{m_1 + m_2}} \begin{pmatrix} 1 \\ 0 \\ 0 \\ i \end{pmatrix} e^{-|m(x)vx|/\hbar + ik_y y} \qquad (2.25)$$

with the dispersion $\epsilon_{k,+} = v\hbar k_y$, and the other has the form

$$\Psi_-(x, k_y) = \sqrt{\frac{v}{h} \frac{m_1 m_2}{m_1 + m_2}} \begin{pmatrix} 0 \\ 1 \\ i \\ 0 \end{pmatrix} e^{-|m(x)vx|/\hbar + ik_y y} \qquad (2.26)$$

with the dispersion $\epsilon_{k,-} = -v\hbar k_y$. We can check these two solutions in this way.
The Dirac equation can be divided into two parts:

$$H = (m(x)v^2 \beta + v p_x \alpha_x) + v p_y \alpha_y. \qquad (2.27)$$

From the one-dimensional solution, we have

$$(m(x)v^2 \beta + v p_x \alpha_x)\Psi_\pm = 0 \qquad (2.28)$$

and

$$v p_y \alpha_y \Psi_\pm = \pm v p_y \Psi_\pm. \qquad (2.29)$$

From the dispersions of the two states, the effective velocities of electrons in the
states are

$$v_\pm = \frac{\partial \epsilon_{k,\pm}}{\hbar \partial k} = \pm v. \qquad (2.30)$$

Therefore, each state carries a current along the interface, but electrons with different spins move in opposite directions. The current density decays exponentially
away from the interface. As the system has time reversal symmetry, the two states
are time reversal counterpart with each other, which constitute a pair of helical edge
(or bound) states at the interface. Furthermore, the Dirac equation of $p_z = 0$ can be
reduced to two independent set of equations:

$$h(x) = v p_x \sigma_x \pm v p_y \sigma_y + m(x)v^2 \sigma_z \qquad (2.31)$$

for different spins. It becomes more clear why two bound states have opposite velocities.

2.2.3 Three and Higher Dimensions

In three and higher dimensions, there always exist the bound states at the interface of the system with positive and negative masses. While all other components of the momentum among the interface are good quantum number, there always exists the solution for zero momentum as in the one-dimensional case. We can use the solutions as the basis to derive the solutions of nonzero momenta in higher dimensions.

2.3 Why Not the Dirac Equation

From the Dirac equation, we have known that there exists a solution of bound states at the interface between two media with positive and negative mass or energy gap. These solutions are quite robust against the roughness of the interface or other factors. If we assume the vacuum is an insulator with an infinite large and positive mass or energy gap, then the system with a negative mass should have bound states around the open boundary. This is very close to the definition of topological insulators. However, because of the symmetry in the Dirac equation with positive and negative masses, there is no topological distinction between these two systems after a unitary transformation. We cannot justify which one is topologically trivial or nontrivial simply from the sign of mass or energy gap. If we use the Dirac equation to describe a topological insulating phase, we have to introduce or assign an additional "vacuum" as a benchmark. Thus, we think this additional condition is unnecessary since the existence of the bound state should have a physical and intrinsic consequence of the band structure in topological insulators. Therefore, we conclude that the Dirac equation in Eq. (2.1) itself may not be a "suitable" candidate to describe the topology of quantum matters.

2.4 Quadratic Correction to the Dirac Equation

To explore possible description to the topological insulator, we introduce a quadratic correction $-B\mathbf{p}^2$ in the momentum $\mathbf{p}$ to the band gap or rest-mass term in the Dirac equation [4]:

$$H = v\mathbf{p} \cdot \alpha + \left(mv^2 - B\mathbf{p}^2\right)\beta, \tag{2.32}$$

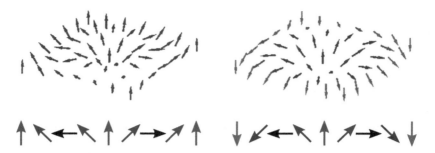

Fig. 2.2 Spin orientation in the momentum space. *Left* ($mB < 0$): the spins at $p = 0$ and $p = +\infty$ are parallel, which is topologically trivial. *Left bottom*: spin orientation along the p_x-axis. *Right* ($mB > 0$): the spins at $p = 0$ and $p = +\infty$ are antiparallel, which is topologically nontrivial. *Right bottom*: spin orientation along the p_x-axis

where mv^2 is the band gap of particle and m and v have dimensions of mass and speed, respectively. B^{-1} also has the dimension of mass. The quadratic term breaks the symmetry between the mass m and $-m$ in the Dirac equation and makes this equation topologically distinct from the original Dirac equation in Eq. (2.1).

To illustrate this, we plot the spin distribution of the ground state in the momentum space, as shown in Fig. 2.2. At $p = 0$, the spin orientation is determined by $mv^2\beta$ or the sign of mass m, but for a large p, it is determined by $-B\mathbf{p}^2\beta$ or sign of B. If the dimensionless parameter $mB > 0$, when p increases along one direction, say the x-direction, the spin will rotate from the z-direction to the x-direction of $\mathbf{p}$ at $p_c^2 = mv^2/B$ and then eventually to the opposite z-direction for a larger $\mathbf{p}$. This consists of two so-called "merons" in the momentum space, which is named skymion. For $mB < 0$, when p increases, the spin will rotate from the z-direction to the direction of p and then flips back to the original z-direction. The feature whether the spin points to the same direction or not at $p = 0$ and $+\infty$ determines the equation topologically distinct in the cases of $mB > 0$ and $mB < 0$.

2.5 Bound State Solutions of the Modified Dirac Equation

2.5.1 One Dimension: End States

Let us start with a one-dimensional case. In this case, the 4×4 Eq. (2.32) can be decoupled into two independent sets of 2×2 equations:

$$h(x) = vp_x\sigma_x + \left(mv^2 - Bp_x^2\right)\sigma_z. \qquad (2.33)$$

For a semi-infinite chain with $x \geq 0$, we consider an open boundary condition at $x = 0$. It is required that the wave function vanishes at the boundary, that is, the Dirichlet boundary condition. Usually we may have a series of solutions of extended

Fig. 2.3 Schematic of the probability density $|\Psi(x)|^2$ of the end state solution as a function of position in Eq. (2.39)

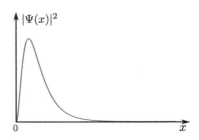

states, which wave functions spread in the whole space. In this section we focus on the solution of the bound state near the boundary. To find the solution of zero energy, we have

$$\left[v\mathbf{p}_x\sigma_x + \left(mv^2 - B\mathbf{p}_x^2\right)\sigma_z\right]\varphi(x) = 0. \tag{2.34}$$

Multiplying σ_x from the left-hand side, one obtains

$$\partial_x\varphi(x) = -\frac{1}{v\hbar}\left(mv^2 + B\hbar^2\partial_x^2\right)\sigma_y\varphi(x). \tag{2.35}$$

If $\varphi(x)$ is an eigenfunction of σ_y, take $\varphi(x) = \chi_\eta\phi(x)$ with $\sigma_y\chi_\eta = \eta\chi_\eta$ $(\eta = \pm 1)$. Then the differential equation is reduced to the second-order ordinary differential equation:

$$\partial_x\phi(x) = -\frac{\eta}{v\hbar}\left(mv^2 + B\hbar^2\partial_x^2\right)\phi(x). \tag{2.36}$$

Taking the trial wave function $\phi(x) \propto e^{-\lambda x}$, one obtains the secular equation

$$B\hbar^2\lambda^2 - \eta v\hbar\lambda + mv^2 = 0. \tag{2.37}$$

The two roots satisfy the relation $\lambda_+ + \lambda_- = \eta v/\hbar B$ and $\lambda_+\lambda_- = mv^2/B\hbar^2$. We require that the wave function vanishes at $x = 0$ and $x = +\infty$:

$$\varphi(x = 0) = \varphi(x = +\infty) = 0. \tag{2.38}$$

The two roots should be positive, and only one of χ_η satisfies the boundary condition for a bound state, $\eta = \text{sgn}(B)$ (without loss of generality, we assume that v is always positive). In the condition of $mB > 0$, there exists a solution of the bound state with zero energy

$$\varphi_\eta(x) = \frac{C}{\sqrt{2}}\begin{pmatrix}\text{sgn}(B)\\i\end{pmatrix}(e^{-x/\xi_+} - e^{-x/\xi_-}) \tag{2.39}$$

where $\xi_\pm^{-1} = \frac{v}{2|B|\hbar}\left(1 \pm \sqrt{1 - 4mB}\right)$ and C is the normalization constant. The main feature of this solution is that the wave function distributes dominantly near the boundary, and decays exponentially away from one end, as shown in Fig. 2.3. The two parameters $\xi_- > \xi_+$ and decide the spatial distribution of the wave function. These are two important length scales, which characterize the end states. When $B \rightarrow 0$, $\xi_+ \rightarrow |B| \hbar / v$ and $\xi_- = \hbar / mv$, that is, ξ_+ approaches to zero, and ξ_- becomes a finite constant which is determined by the energy gap mv^2. If we relax the constraint of the vanishing wave function at the boundary, the solution exists even if $B = 0$. In this way, we go back the conventional Dirac equation. In this sense, the two equations reach at the same conclusion. When $m \rightarrow 0$, $\xi_- = \hbar / mv \rightarrow +\infty$, and the state evolves into a bulk state. Thus, the end states disappear and a topological quantum phase transition occurs at $m = 0$.

In the four-component form to Eq. (2.32), two degenerate solutions have the form

$$\Psi_1 = \frac{C}{\sqrt{2}} \begin{pmatrix} \mathrm{sgn}(B) \\ 0 \\ 0 \\ i \end{pmatrix} (\mathrm{e}^{-x/\xi_+} - \mathrm{e}^{-x/\xi_-}), \tag{2.40}$$

$$\Psi_2 = \frac{C}{\sqrt{2}} \begin{pmatrix} 0 \\ \mathrm{sgn}(B) \\ i \\ 0 \end{pmatrix} (\mathrm{e}^{-x/\xi_+} - \mathrm{e}^{-x/\xi_-}). \tag{2.41}$$

We shall see that these two solutions can be used to derive the effective Hamiltonian for higher dimensional systems.

The role of this solution could not be underestimated in theory of topological insulators. We shall see that all other solutions of the edge or surface states and topological excitations can be closely related to this solution.

2.5.2 Two Dimensions: Helical Edge States

In two dimensions, the equation can be also decoupled into two independent equations:

$$h_\pm = v p_x \sigma_x \pm v p_y \sigma_y + \left(mv^2 - B p^2 \right) \sigma_z. \tag{2.42}$$

These two equations break the "time" reversal symmetry under the transformation of $\sigma_i \rightarrow -\sigma_i$ and $p_i \rightarrow -p_i$, although the original four-component equation is time reversal invariant.

We consider a semi-infinite plane with the boundary at $x = 0$. $p_y = \hbar k_y$ is a good quantum number. At $k_y = 0$, the two-dimensional equation has the identical form as the one-dimensional equation. The x-dependent part of the solution has the identical form as that in one dimension. Thus, we use the two one-dimensional solutions $\{\Psi_1, \Psi_2\}$ in Eqs. (2.40) and (2.41) as the basis. The y-dependent part $\Delta H_{2D} = v p_y \alpha_y - B p_y^2 \beta$ is regarded as the perturbation to the one-dimensional Hamiltonian. In this way, we have a one-dimensional effective model for the helical edge states:

$$H_{\text{eff}} = ((\langle\Psi_1|, \langle\Psi_2|)\Delta H_{2D} \begin{pmatrix} |\Psi_1\rangle \\ |\Psi_2\rangle \end{pmatrix} = vp_y\text{sgn}(B)\sigma_z. \tag{2.43}$$

The sign dependence of B in the effective model also reflects the fact that the helical edge states disappear if $B = 0$. The dispersion relations for the bound states at the boundary are

$$\epsilon_{p_y,\pm} = \pm vp_y. \tag{2.44}$$

Electrons have positive $(+v)$ and negative velocity $(-v)$ in two different states, respectively, and form a pair of helical edge states.

The exact solutions of the edge states to this two-dimensional equation have the similar form of one-dimensional equation [5]:

$$\Psi_1 = \frac{C}{\sqrt{2}} \begin{pmatrix} \text{sgn}(B) \\ 0 \\ 0 \\ i \end{pmatrix} (e^{-x/\xi_+} - e^{-x/\xi_-})e^{+ip_yy/\hbar}, \tag{2.45a}$$

$$\Psi_2 = \frac{C}{\sqrt{2}} \begin{pmatrix} 0 \\ \text{sgn}(B) \\ i \\ 0 \end{pmatrix} (e^{-x/\xi_+} - e^{-x/\xi_-})e^{+ip_yy/\hbar}, \tag{2.45b}$$

with the dispersion relations $\epsilon_{p_y,\pm} = \pm vp_y\text{sgn}(B)$. The characteristic lengths become p_y dependent,

$$\xi_\pm^{-1} = \frac{v}{2|B|\hbar}\left(1 \pm \sqrt{1 - 4mB + 4B^2p_y^2/v^2}\right). \tag{2.46}$$

In two dimensions, the Chern number or Thouless-Kohmoto-Nightingale-Nijs (TKNN) integer can be used to characterize whether the system is topologically trivial or nontrivial [6]. For the two-band Hamiltonian in the form $H = \mathbf{d}(p) \cdot \sigma$, the Chern number is expressed as

$$n_c = -\frac{1}{4\pi}\int d\mathbf{p}\frac{\mathbf{d} \cdot (\partial_{p_x}\mathbf{d} \times \partial_{p_y}\mathbf{d})}{d^3} \tag{2.47}$$

where $d^2 = \sum_{\alpha=x,y,z} d_\alpha^2$ (see Appendix A.2). The integral runs over the first Brillouin zone for a lattice system, in which the number n_c is always an integer (see Appendix A.1). In the continuous limit, the integral area becomes infinite; the integral can be fractional. For Eq. (2.42), the Chern number has the form [7, 8]

$$n_\pm = \pm(\text{sgn}(m) + \text{sgn}(B))/2. \tag{2.48}$$

which is related to the Hall conductance $\sigma_{\pm} = n_{\pm}e^2/h$. When m and B have the same sign, $n_{\pm} = \pm 1$, and the system is topologically nontrivial. But if m and B have different signs, $n_{\pm} = 0$. The topologically nontrivial condition is in agreement with the existence condition of edge-state solution $mB > 0$. This reflects the bulk-edge relation of the integer quantum Hall effect [9].

2.5.3 Three Dimensions: Surface States

In three dimensions, we consider a $y - z$ plane at $x = 0$. We can derive an effective model for the surface states by means of the one-dimensional solution of the bound state. Since the momenta among the $y - z$ plane are good quantum numbers, we use their eigenvalues to replace the momentum operators, p_y and p_z. Consider p_y- and p_z-dependent part as a perturbation to $H_{1D}(x)$:

$$\Delta H_{3D} = v p_y \alpha_y + v p_z \alpha_z - B(p_y^2 + p_z^2)\beta. \tag{2.49}$$

The solutions of the three-dimensional Dirac equation at $p_y = p_z = 0$ are identical to two one-dimensional solutions, $|\Psi_1\rangle$ and $|\Psi_2\rangle$ in Eqs. (2.40) and (2.41). For $p_y, p_z \neq 0$, we use the solution

$$\Psi_1 = \frac{C}{\sqrt{2}} \begin{pmatrix} \mathrm{sgn}(B) \\ 0 \\ 0 \\ i \end{pmatrix} (e^{-x/\xi_+} - e^{-x/\xi_-}) e^{i(p_y y + p_z z)/\hbar}, \tag{2.50a}$$

$$\Psi_2 = \frac{C}{\sqrt{2}} \begin{pmatrix} 0 \\ \mathrm{sgn}(B) \\ i \\ 0 \end{pmatrix} (e^{-x/\xi_+} - e^{-x/\xi_-}) e^{i(p_y y + p_z z)/\hbar} \tag{2.50b}$$

as the basis. A straightforward calculation as in the two-dimensional case gives

$$H_{\mathrm{eff}} = (\langle\Psi_1|, \langle\Psi_2|)\Delta H_{3D} \begin{pmatrix} |\Psi_1\rangle \\ |\Psi_2\rangle \end{pmatrix} = v\,\mathrm{sgn}(B)(p \times \sigma)_x. \tag{2.51}$$

Under a unitary transformation,

$$\Phi_1 = \frac{1}{\sqrt{2}}(|\Psi_1\rangle - i\,|\Psi_2\rangle), \tag{2.52a}$$

$$\Phi_2 = \frac{-i}{\sqrt{2}}(|\Psi_1\rangle + i\,|\Psi_2\rangle), \tag{2.52b}$$

Fig. 2.4 The Dirac cone of
the surface states in the
momentum space

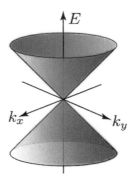

one can have a gapless Dirac equation for the surface states

$$H_{\text{eff}} = (\langle\Phi_1|, \langle\Phi_2|)\Delta H_{3D} \begin{pmatrix} |\Phi_1\rangle \\ |\Phi_2\rangle \end{pmatrix}$$

$$= v\text{sgn}(B)(p_y\sigma_y + p_z\sigma_z). \tag{2.53}$$

The dispersion relations become $\epsilon_{p,\pm} = \pm vp$ with $p = \sqrt{p_y^2 + p_z^2}$. In this way we have an effective model for a single Dirac cone of the surface states, as plotted in Fig. 2.4. Note that σ_i in the Hamiltonian is not a real spin, which is determined by two states at $p_y = p_z = 0$. In some systems, $|\Psi_1\rangle$ and $|\Psi_2\rangle$ are almost polarized along the z-direction of electron spin. In this sense the Pauli matrices in Eq. (2.51) may be regarded approximately as a real spin.

The exact solutions of the surface states to this three-dimensional equation with the boundary are

$$\Psi_\pm = C\Psi_\pm^0(e^{-x/\xi_+} - e^{-x/\xi_-})\exp[+i\left(p_y y + p_z z\right)/\hbar] \tag{2.54}$$

where

$$\Psi_+^0 = \begin{pmatrix} \cos\frac{\theta}{2}\text{sgn}(B) \\ -i\sin\frac{\theta}{2}\text{sgn}(B) \\ \sin\frac{\theta}{2} \\ i\cos\frac{\theta}{2} \end{pmatrix}, \tag{2.55a}$$

$$\Psi_-^0 = \begin{pmatrix} \sin\frac{\theta}{2}\text{sgn}(B) \\ i\cos\frac{\theta}{2}\text{sgn}(B) \\ -\cos\frac{\theta}{2} \\ i\sin\frac{\theta}{2} \end{pmatrix} \tag{2.55b}$$

with the dispersion relation $\epsilon_{p,\pm} = \pm v p \text{sgn}(B)$. $\tan\theta = p_y/p_z$. The penetration depth becomes p dependent,

$$\xi_{\pm}^{-1} = \frac{v}{2\,|B|\,\hbar}\left(1 \pm \sqrt{1 - 4mB + 4B^2 p^2/\hbar^2}\right). \qquad (2.56)$$

2.5.4 Generalization to Higher-Dimensional Topological Insulators

The solution can be generalized to higher-dimensional systems. We conclude that there always exist d-dimensional boundary or surface states in the (d+1)-dimensional modified Dirac equation when $mB > 0$.

2.6 Summary

From the solutions of the modified Dirac equation, we found that under the condition of $mB > 0$,

- In one dimension, there exists bound state of zero energy near the ends.
- In two dimensions, there exists helical edge states near the edge.
- In three dimensions, there exists surface states near the surface.
- In higher dimensions, there always exists higher dimensional boundary state.

From the solutions of the bound states near the boundary and the calculation of Z_2 index, we conclude that the modified Dirac equation can provide a description of a large class of topological insulators from one to higher dimensions.

2.7 Further Reading

- J.D. Bjorken, S.D. Drell, *Relativistic Quantum Mechanics* (MaGraw-Hill, New York, 1964)
- P.A.M. Dirac, *Principles of Quantum Mechanics*, 4th edn. (Clarendon, Oxford, 1982)
- S.Q. Shen, W.Y. Shan, H.Z. Lu, Topological insulator and the Dirac equation. SPIN **1**, 33 (2011)

References

1. P.A.M. Dirac, Proc. R. Soc. A **117**, 610 (1928)
2. P.A.M. Dirac, *Principles of Quantum Mechanics*, 4th edn. (Clarendon, Oxford, 1982)

3. R. Jackiw, C. Rebbi, Phys. Rev. D **13**, 3398 (1976)
4. S.Q. Shen, W.Y. Shan, H.Z. Lu. SPIN **1**, 33 (2011)
5. B. Zhou, H.Z. Lu, R.L. Chu, S.Q. Shen, Q. Niu, Phys. Rev. Lett. **101**, 246807 (2008)
6. D.J. Thouless, M. Kohmoto, M.P. Nightingale, M. den Nijs, Phys. Rev. Lett. **49**, 405 (1982)
7. H.Z. Lu, W.Y. Shan, W. Yao, Q. Niu, S.Q. Shen, Phys. Rev. B **81**, 115407 (2010)
8. W.Y. Shan, H.Z. Lu, S.Q. Shen, New J. Phys. **12**, 043048 (2010)
9. Y. Hatsugai, Phys. Rev. Lett. **71**, 3697 (1993)

Chapter 3
Minimal Lattice Model for Topological Insulator

Abstract A lattice model can be mapped into a continuous one near the critical point of topological quantum phase transition. Topology of a lattice model remains unchanged if no energy gap in the band structure closes and reopens.

Keywords The lattice model • Band gap • Parity • Time reversal invariant momentum

3.1 Tight Binding Approximation

A tight binding model is extensively used to describe the band structure of electrons in solids. A schematic in Fig. 3.1 depicts the formation of the tight binding lattice from the point of view of atom physics. Consider an isolated atom, say the hydrogen atom. In quantum mechanics, an electron rotates around the nuclei in the Coulomb interaction and forms a series of discrete energy levels or orbits, $E_n = -e^2/(8\pi\epsilon_0 n^2 a_0)$ where $a_0 = 4\pi\epsilon_0\hbar^2/(m_e e^2)$ is the Bohr radius and n is an integer. The ground state energy is $E_{n=1} = -13.6\,\text{eV}$, and the radius of the orbit is $a_0 = 0.529\text{Å}$. The energy of the first excited state is $E_{n=2} = -3.4\,\text{eV}$. The energy difference between the two states is about $-10.2\,\text{eV}$, which is very large in a solid. Thus, it is a good approximation to consider only the ground state of electron at low temperatures. When two atoms get closer, the orbits of two electrons of different atoms may overlap in space. As a result, electron of one atom has a probability to jump into the orbit of another atom. Since the electron is mainly localized around the original nuclei, the probability of the electron tunneling from one atom to another is still quite tiny. The picture can be generalized to a lattice system consisting of atoms: the electrons move from one atom to another one and form energy bands.

S.-Q. Shen, *Topological Insulators: Dirac Equation in Condensed Matters*,
Springer Series in Solid-State Sciences 174, DOI 10.1007/978-3-642-32858-9_3,
© Springer-Verlag Berlin Heidelberg 2012

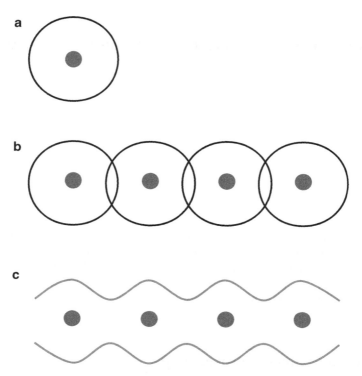

Fig. 3.1 A schematic which explains the tight binding approximation. (**a**) A single atom with discrete orbits for electron. (**b**) When atoms get together to form a solid, the wave functions of two orbits (*black*) of adjunct atoms overlap in space. (**c**) If the overlap of the orbits is small, the electrons are still regarded to be almost localized around the original orbits but have a tiny probability to tunnel to the adjunct orbits to form an energy band

In the second quantization, an effective model Hamiltonian is written as

$$H = \sum_{i,\sigma=\uparrow,\downarrow} \epsilon_0 c_{i,\sigma}^\dagger c_{i,\sigma} - \sum_{\langle i,j\rangle,\sigma=\uparrow,\downarrow} t_{ij} c_{i,\sigma}^\dagger c_{j,\sigma}, \tag{3.1}$$

where the summation i runs over all the lattice sites and $\sigma =\uparrow,\downarrow$ represents for the electron spin-up and spin-down, respectively. $c_{i,\sigma}^\dagger$ and $c_{i,\sigma}$ are the creation and annihilation operators of electron with spin σ at site i, respectively, obeying the anticommutation relation, $c_{i,\sigma}^\dagger c_{j,\sigma'} + c_{j,\sigma'} c_{i,\sigma}^\dagger = \delta_{\sigma\sigma'}\delta_{ij}$. It is required that $c_{i,\sigma}|0\rangle = 0$. t_{ij} describes the hopping amplitude of electron jumping from site i to site j.

For a ring of one-dimensional lattice with N lattice sites or a one-dimensional lattice with a periodic boundary condition, we take $c_{i,\sigma} = c_{i+N,\sigma}$. For simplicity, we just suppose the lattice is translationally invariant by taking $t_{ij} = t$ for a pair of nearest neighbor lattice sites. Performing the Fourier transformation,

$$c_{i,\sigma} = \frac{1}{\sqrt{Na}} \sum_{k_n} e^{ik_n R_i} c_{k_n,\sigma}, \tag{3.2a}$$

$$c_{i,\sigma}^\dagger = \frac{1}{\sqrt{Na}} \sum_{k_n} e^{-ik_n R_i} c_{k_n,\sigma}^\dagger, \tag{3.2b}$$

and the periodic boundary condition gives $e^{ik_n R_i} = e^{ik_n(R_i+Na)}$ and $k_n = 2n\pi/Na$ ($n = 0, 1, \ldots, N-1$). In this way, the Hamiltonian can be diagonalized

$$H = \sum_{k_n} \epsilon(k_n) c_{k_n,\sigma}^\dagger c_{k_n,\sigma} \tag{3.3}$$

with the dispersion $\epsilon(k_n) = \epsilon_0 - 2t \cos k_n a$. Notice that for $K = 2\pi/a$, $\epsilon(k_n+K) = \epsilon(k_n)$. For a very large N, $k_n a$ can be taken to be continuous from 0 to 2π, and K is called the reciprocal lattice vector.

The approach can be generalized to two and three dimensions. The reciprocal lattice vector can be defined for the purpose of the Fourier transformation from the real space to the momentum space. In a three-dimensional lattice with the lattice space **a**, **b**, and **c**, the reciprocal lattice vectors are given by

$$\mathbf{K}_a = 2\pi \frac{\mathbf{b} \times \mathbf{c}}{\mathbf{a} \cdot (\mathbf{b} \times \mathbf{c})}, \tag{3.4a}$$

$$\mathbf{K}_b = 2\pi \frac{\mathbf{c} \times \mathbf{a}}{\mathbf{a} \cdot (\mathbf{b} \times \mathbf{c})}, \tag{3.4b}$$

$$\mathbf{K}_c = 2\pi \frac{\mathbf{a} \times \mathbf{b}}{\mathbf{a} \cdot (\mathbf{b} \times \mathbf{c})}, \tag{3.4c}$$

and $\mathbf{K}_\alpha \cdot \mathbf{R}_\alpha = 2\pi$ for $\alpha = a, b, c$.

3.2   From Continuous to Lattice Model

Usually a continuous model is for low-energy physics in the long wavelength limit. Topology of the band structure should reveal the properties of the whole band structure in the Brillouin zone. In practice, people like to use a lattice model instead of a continuous model to explore the topology of system. A continuous model can be mapped into a lattice model in the tight binding approximation, in which the Brillouin zone is finite. In a d-dimensional hypercubic lattice, one replaces [1],

$$k_i \rightarrow \frac{1}{a} \sin k_i a, \tag{3.5a}$$

$$k_i^2 \rightarrow \frac{4}{a^2} \sin^2 \frac{k_i a}{2} = \frac{2}{a^2}(1 - \cos k_i a), \tag{3.5b}$$

which are equal only in the long wavelength limit, that is, $k_i a \rightarrow 0$ by using the relation $\sin x \approx x$ for a small x. We use $\sin^2 \frac{k_i a}{2}$ or $\cos k_i a$ instead of $\sin^2 k_i a$ for k_i^2 to avoid long-distance hopping in the effective Hamiltonian. In this way, the hopping terms in the lattice model only exist between the nearest neighbor sites.

Usually there exits the fermion doubling problem in the lattice model for massless Dirac particles. The replacement of $k_i \rightarrow \frac{1}{a} \sin k_i a$ will cause an additional zero point for $\frac{1}{a} \sin k_i a$ at $k_i a = \pi$ besides at $k_i a = 0$. Thus, there exist four Dirac cones in a square lattice at $k = (0,0)$, $(0, \pi/a)$, $(\pi/a, 0)$ and $(\pi/a, \pi/a)$, for a gapless Dirac equation. A large B term removes the zero point at $(\pi/a, \pi/a)$ as $\frac{4B}{a^2} \sin^2 \frac{k_i a}{2} \rightarrow Bk^2$. Thus, the lattice model is equivalent to the continuous model only in the condition of a large B. Thus, for a finite B, the band gap may not open at the Γ point in the lattice model because of the competition between the linear term and the quadratic term of k_i. This fact may lead to a topological transition from a large B to a small B. Imura et al. [2] analyzed the two-dimensional case in details and found that there exists a topological transition at a finite value of B in two dimensions. A similar transition will also exist in higher dimensions. Thus, it should be careful when we study the continuous model in a tight binding approximation. However, the topology of the band structure never changes if the energy gap in the band structure does not close and reopen while the model parameters vary continuously.

With this mapping, one obtains a lattice model for topological insulator

$$H = \frac{\hbar v}{a} \sum_{i=x,y,z} \sin k_i a \alpha_i + \left(mv^2 - B \frac{4\hbar^2}{a^2} \sum_{i=x,y,z} \sin^2 \frac{k_i a}{2} \right) \beta. \tag{3.6}$$

The energy dispersions for this system are

$$E_{k,\pm} = \pm \sqrt{\frac{\hbar^2 v^2}{a^2} \sum_{i=x,y,z} \sin^2 k_i a + \left(mv^2 - \frac{4B\hbar^2}{a^2} \sum_{i=x,y,z} \sin^2 \frac{k_i a}{2} \right)^2}. \tag{3.7}$$

For $mB < 0$, there is always an energy gap between the two bands $2|m|v^2$. For $mB = 0$ ($B \neq 0$), the energy gap closes at the points $k_i a = 0$ as $E_{0,+} = E_{0,-}$. For $mB > 0$, there exist several gapless points at $mv^2 = 4B\hbar^2/a^2$ (in one, two, and three dimensions), $8B\hbar^2/a^2$ (in two and three dimensions), and $12B\hbar^2/a^2$ (in three dimensions). We shall show that these are the critical points for topologically quantum phase transition. For simplicity, we take the lattice constant $a = \hbar = 1$.

We can perform the Fourier transform to transfer the effective Hamiltonian from the momentum space into the lattice space. In the tight binding approximation, the model Hamiltonian on a hypercubic lattice has the form

$$H = \sum_{i,n,m} \Delta c_{i,n}^{\dagger} \beta_{nm} c_{i,m} - t \sum_{\langle i,j \rangle,} c_{j,n}^{\dagger} \beta_{nm} c_{i,m}$$

$$+ it' \sum_{i,a,n,m} \left[c_{i+a,n}^{\dagger} (\alpha_a)_{nm} c_{i,m} - c_{i,n}^{\dagger} (\alpha_a)_{nm} c_{i+a,m} \right]. \tag{3.8}$$

Here, $\langle i, j \rangle$ runs over the pairs of nearest neighbor sites. $a = x, y, z$ and $i + a$ represents the lattice site $R_i + R_a$. $n, m = 1, 2, \ldots, D$, where D is the dimension of the Dirac matrices. The relations of the model parameters are

$$t' = \frac{\hbar v}{2a} = v/2, \ \Delta - 2dt = mv^2, \ t = -B\hbar^2/a^2 = -B. \tag{3.9}$$

Denote $(c_{i,1}^{\dagger}, c_{i,2}^{\dagger}, \ldots, c_{i,D}^{\dagger})$ by $c_i^{\dagger}$. In this way, the equation can be written in a compact form

$$H = \sum_i \Delta c_i^{\dagger} \beta c_i - t \sum_{\langle i,j \rangle,} c_j^{\dagger} \beta c_i + it' \sum_{i,a} \left[c_{i+a}^{\dagger} \alpha_a c_i - c_i^{\dagger} \alpha_a c_{i+a} \right]. \tag{3.10}$$

3.3 One-Dimensional Lattice Model

Consider a one-dimensional lattice model:

$$H = \Delta \sum_{j=1}^{N} c_j^{\dagger} \sigma_z c_j - t \sum_{j=1}^{N-1} \left(c_{j+1}^{\dagger} \sigma_z c_j + c_j^{\dagger} \sigma_z c_{j+1} \right)$$

$$+ it' \sum_{j=1}^{N-1} \left(c_{j+1}^{\dagger} \sigma_x c_j - c_j^{\dagger} \sigma_x c_{j+1} \right), \tag{3.11}$$

where $c_j^{\dagger} = (c_{j,\uparrow}^{\dagger}, c_{j,\downarrow}^{\dagger})$. To find the end state, we adopt the open boundary condition. We choose $\left(c_1^{\dagger}, c_2^{\dagger}, \ldots, c_N^{\dagger} \right)$ as the basis. The Hamiltonian can be written in the form of matrix,

$$H = \begin{pmatrix} \Delta\sigma_z & T & 0 & 0 & \cdots & 0 \\ T^\dagger & \Delta\sigma_z & T & 0 & \cdots & 0 \\ 0 & T^\dagger & \Delta\sigma_z & T & \cdots & 0 \\ \vdots & \vdots & \ddots & \ddots & \ddots & \vdots \\ 0 & 0 & 0 & T^\dagger & \Delta\sigma_z & T \\ 0 & 0 & 0 & 0 & T^\dagger & \Delta\sigma_z \end{pmatrix}, \tag{3.12}$$

where $T = -t\sigma_z - it'\sigma_x$. Since σ_x and σ_z are 2×2 matrices, the Hamiltonian is a $2N \times 2N$ square matrix.

Here, we present a solution for $N = +\infty$, that is, a semi-infinite chain with an end at $j = 1$. We take the eigenvector for H as $\Psi^\dagger = (\Psi_1^\dagger, \Psi_2^\dagger, \ldots, \Psi_N^\dagger)$. The eigenvalue equation of this problem becomes

$$\Delta\sigma_z\Psi_j + T\Psi_{j+1} + T^+\Psi_{j-1} = E\Psi_j \tag{3.13}$$

for $j = 1, 2, \ldots$ and $\Psi_0 = 0$. To solve this equation, we set a trial solution,

$$\Psi_{j+1} = \lambda\Psi_j = \lambda^{j+1}\Psi. \tag{3.14}$$

Then Eq. (3.13) becomes

$$\left(\Delta\sigma_z + \lambda T + \lambda^{-1}T^+\right)\Psi = P\Psi = E\Psi, \tag{3.15}$$

where the operator $P = \Delta\sigma_z + \lambda T + \lambda^{-1}T^+ \equiv \gamma \cdot \sigma$ with

$$\gamma_x = -it'(\lambda - \lambda^{-1}), \tag{3.16a}$$

$$\gamma_y = 0, \tag{3.16b}$$

$$\gamma_z = \Delta - \lambda t - t\lambda^{-1}. \tag{3.16c}$$

In general, the matrix P is non-Hermitian, and one may have two complex eigenvalues for P. However, E must be real as the eigenvalue for a physical system. Thus, P should meet one of the conditions:

1. All components of γ are real.
2. All nonzero complex components combine to give $E = 0$.

The first condition is met when $\gamma = e^{ik}$, which gives the solution of the bulk band. These solutions are not what we are interested in here. The second condition defines the so-called annihilator. In the present case, if $\gamma_z = is\gamma_x$ ($s = \pm1$),

$$P = \gamma_x(\sigma_x + is\sigma_z). \tag{3.17}$$

$\Psi = \frac{1}{\sqrt{2}}(1, -is)^T$ satisfies $P\Psi = 0$, which is also one of the eigenstates of σ_y with the eigenvalue $-s$.

Increasing and decreasing operators are defined by $\sigma_\pm = \sigma_x \pm i\sigma_y$, which satisfy

$$\sigma_+ \begin{pmatrix} 1 \\ 0 \end{pmatrix} = 0; \sigma_+ \begin{pmatrix} 0 \\ 1 \end{pmatrix} = 2 \begin{pmatrix} 1 \\ 0 \end{pmatrix}, \tag{3.18a}$$

$$\sigma_- \begin{pmatrix} 0 \\ 1 \end{pmatrix} = 0; \sigma_- \begin{pmatrix} 1 \\ 0 \end{pmatrix} = 2 \begin{pmatrix} 0 \\ 1 \end{pmatrix}. \tag{3.18b}$$

To have a zero-energy mode of $E = 0$, one has

$$\Delta - \lambda t - t\lambda^{-1} = st'(\lambda - \lambda^{-1}). \tag{3.19}$$

This equation has two roots:

$$\lambda_\pm(s) = \frac{\Delta}{2(t + st')} \left[1 \pm \sqrt{1 - \frac{4(t^2 - t'^2)}{\Delta^2}} \right]. \tag{3.20}$$

The solutions for the end state require $|\lambda_\pm| < 1$ as $\Psi_j \to 0$ for a large j. Thus,

$$\lambda_+\lambda_- = \frac{t - st'}{t + st'} < 1 \tag{3.21}$$

which requires $s = \text{sgn}(t'/t)$.

Case I: $\lambda_\pm$ are complex for

$$4(t^2 - t'^2) > \Delta^2 \tag{3.22}$$

and

$$|\lambda_+^2| = \frac{t - st'}{t + st'} = \frac{1 - \left|\frac{t'}{t}\right|}{1 + \left|\frac{t'}{t}\right|}. \tag{3.23}$$

Case II: For

$$4(t^2 - t'^2) < \Delta^2, \tag{3.24}$$

we require

$$|\lambda_\pm|^2 = \frac{\Delta^2}{4(t + st')^2} \left[2 - \frac{4(t^2 - t'^2)}{\Delta^2} \pm 2\sqrt{1 - \frac{4(t^2 - t'^2)}{\Delta^2}} \right] < 1. \tag{3.25}$$

It follows that

$$4(t^2 - t'^2) < \Delta^2 < 4t^2. \tag{3.26}$$

Thus, the boundary condition of $\Psi_0 = 0$ gives the solution

$$\Psi_j = (\lambda_+^j - \lambda_-^j)\Psi, \qquad (3.27)$$

which does not vanish at the boundary at $j = 1$.

A special case: there are two exact solutions of this lattice model at $\Delta = 0$ and $t = t'$. In this case, we have the solutions

$$\Psi_L = \begin{pmatrix} \varphi_1 \\ 0 \\ \vdots \\ 0 \end{pmatrix}, \Psi_R = \begin{pmatrix} 0 \\ 0 \\ \vdots \\ \varphi_N \end{pmatrix}, \qquad (3.28)$$

with

$$T^\dagger \varphi_1 = -t(\sigma_z - i\sigma_x)\varphi_1 = 0, \qquad (3.29a)$$

$$T\varphi_N = -t(\sigma_z + i\sigma_x)\varphi_N = 0, \qquad (3.29b)$$

and

$$\varphi_1 = \frac{1}{\sqrt{2}}\begin{pmatrix} 1 \\ -i \end{pmatrix}, \varphi_N = \frac{1}{\sqrt{2}}\begin{pmatrix} 1 \\ +i \end{pmatrix}. \qquad (3.30)$$

These two solutions are located at two ends, and the energy eigenvalues are 0. As the two solutions are degenerate, a linear combination of these two solutions is also the solution for the lattice model.

3.4 Two-Dimensional Lattice Model

3.4.1 Integer Quantum Hall Effect

In two dimensions, the lattice model on a square lattice can be written as

$$H = \mathbf{d}(\mathbf{k}) \cdot \sigma \qquad (3.31)$$

where

$$d_x = A \sin k_x a, \qquad (3.32a)$$

$$d_y = A \sin k_y a, \qquad (3.32b)$$

$$d_z = \Delta - 4B \sin^2 \frac{k_x a}{2} - 4B \sin^2 \frac{k_y a}{2}. \qquad (3.32c)$$

One can regard this model as a quantum spin-$\frac{1}{2}$ in an effective magnetic field, $\mathbf{d}(k)$. The dispersion relations are

$$E_{k,\pm} = \pm |\mathbf{d}(\mathbf{k})|. \tag{3.33}$$

The zero points of the dispersion are determined by a set of equations

$$\sin^2 k_x a = \sin^2 k_x a = 0, \tag{3.34a}$$

$$\Delta = 4B \sin^2 \frac{k_x a}{2} + 4B \sin^2 \frac{k_y a}{2}. \tag{3.34b}$$

There are three solutions: (1) $\Delta = 0$ with ($k_x a = 0$, $k_y a = 0$); (2) $\Delta = 4B$ with ($k_x a = 0$, $k_y a = \pi$) or ($k_x a = \pi$, $k_y a = 0$); and (3) $\Delta = 8B$ with ($k_x a = \pi$, $k_y a = \pi$). Thus, the energy gap closes and reopens near these points. We shall see that topological quantum phase transition will occur at the points (1) $\Delta = 0$, (2) $\Delta = 4B$, and (3) $\Delta = 8B$.

To have a solution of edge state, we may take a geometry of ribbon. Along the x-direction, we adopt the periodic boundary condition such that k_x is a good quantum number. Along the y-direction, we adopt an open boundary condition. Performing the partial Fourier transformation only for the x-direction, the problem is reduced to a one-dimensional one as k_x is regarded as a variable.

$$H(k_x) = \sum_{j=1}^{N} c_{k_x,j}^{\dagger} h_{j,j}(k_x) c_{k_x,j}$$

$$+ \sum_{j=1}^{N-1} \left[c_{k_x,j}^{\dagger} h_{j,j+1}(k_x) c_{k_x,j+1} + c_{k_x,j+1}^{\dagger} h_{j+1,j}(k_x) c_{k_x,j} \right], \tag{3.35}$$

where

$$h_{j,j}(k_x) = A \sin k_x \sigma_x + \left(\Delta - 2B - 4B \sin^2 \frac{k_x a}{2} \right) \sigma_z, \tag{3.36a}$$

$$h_{j,j+1}(k_x) = B\sigma_z - \frac{i}{2} A\sigma_y, \tag{3.36b}$$

$$h_{j+1,j}(k_x) = h_{j,j+1}^{\dagger}(k_x) = B\sigma_z + \frac{i}{2} A\sigma_y. \tag{3.36c}$$

The problem to find the solution of the edge state is reduced to one as in one dimension for a specific k_x. One can solve it following the method introduced in the preceding section. This model can also be solved numerically.

3.4.2 Quantum Spin Hall Effect

Combination of two 2×2 modified Dirac models can generate an effective model for the quantum spin Hall effect. Under the time reversal $\Theta = i\sigma_y K$,

$$k_i \longrightarrow -k_i, \sigma_i \longrightarrow -\sigma_i, \tag{3.37}$$

we have

$$
\begin{aligned}
\Theta \mathbf{d}(-\mathbf{k}) \cdot \sigma \Theta^{-1} &= -\mathbf{d}(-\mathbf{k}) \cdot \sigma \\
&= A\sigma_x \sin k_x a + A\sigma_y \sin k_y a \\
&\quad - \sigma_z \left(\Delta - 4B \sin^2 \frac{k_x a}{2} - 4B \sin^2 \frac{k_y a}{2} \right).
\end{aligned}
\tag{3.38}
$$

We set $\mathbf{d}(\mathbf{k}) \cdot \sigma$ for "spin-up" sector and then $-\mathbf{d}(-\mathbf{k}) \cdot \sigma$ for "spin-down" sector. In this way, we obtain an effective Hamiltonian in a 4×4 matrix:

$$
\begin{aligned}
H_{QSHE} &= \begin{pmatrix} \mathbf{d}(\mathbf{k}) \cdot \sigma & 0 \\ 0 & -\mathbf{d}(-\mathbf{k}) \cdot \sigma \end{pmatrix} \\
&= A \sin k_x a \sigma_x \otimes s_0 + A \sin k_y a \sigma_y \otimes s_0 \\
&\quad + \left(\Delta - 4B \sin^2 \frac{k_x a}{2} - 4B \sin^2 \frac{k_y a}{2} \right) \sigma_z \otimes s_z,
\end{aligned}
\tag{3.39}
$$

where s_0 is a 2×2 identity matrix and s_z is the Pauli matrix for spin index. More terms can be introduced, such as the spin-orbit coupling which appears as an off-diagonal term in the matrix to couple the spin-up and spin-down. In this way, S_z is no longer conserved. However, the edge states may persist. This can be checked numerically.

3.5 Three-Dimensional Lattice Model

The lattice model on a cubic lattice is

$$H = A \sum_{i=x,y,z} \alpha_i \sin k_i a + \beta \left(\Delta - 4B \sum_{i=x,y,z} \sin^2 \frac{k_i a}{2} \right). \tag{3.40}$$

Its dispersions are

$$E_{k,\pm} = \pm |\mathbf{d}(\mathbf{k})|$$

$$= \pm \sqrt{A^2 \sum_{i=x,y,z} \sin^2 k_i a + \left(\Delta - 4B \sum_{i=x,y,z} \sin^2 \frac{k_i a}{2}\right)^2}. \qquad (3.41)$$

The zero points of the dispersion are determined by a set of equations

$$\sin^2 k_x a = \sin^2 k_y a = \sin^2 k_z a = 0, \qquad (3.42a)$$

$$\Delta = 4B \sin^2 \frac{k_x a}{2} + 4B \sin^2 \frac{k_y a}{2} + 4B \sin^2 \frac{k_z a}{2}. \qquad (3.42b)$$

There are three solutions at $\Delta = 0$, $\Delta = 4B$, $\Delta = 8B$, and $\Delta = 12B$. The topological nontrivial regions are $0 < \Delta/B < 4$ and $8 < \Delta/B < 12$. Topological quantum phase transition occurs at the points of $\Delta = 0$ and $\Delta/B = 4, 8$, and 12.

To find the solution of the surface states, we consider a semi-infinite x-y plane. In this case, the k_x and k_y are still good quantum numbers. In this case, performing the partial Fourier transformation for the x- and y-axis,

$$c_{k_x,k_y,j_z} = \frac{1}{\sqrt{N_x N_y}} \sum_{j_x,j_y} c_{j_x,j_y,j_z} \exp[i(k_x j_x + k_y j_y)], \qquad (3.43a)$$

$$c_{j_x,j_y,j_z} = \frac{1}{\sqrt{N_x N_y}} \sum_{k_x,k_y} c_{k_x,k_y,j_z} \exp[-i(k_x j_x + k_y j_y)], \qquad (3.43b)$$

we have a one-dimensional effective Hamiltonian along the z-axis

$$H(k_x, k_y) = \sum_i c^\dagger_{k_x,k_y,j_z} \epsilon(k_x, k_y) c_{k_x,k_y,j_z}$$

$$+ \sum_i c^\dagger_{k_x,k_y,j_z+1} (i\frac{A}{2}\alpha_z - 2B\beta) c_{k_x,k_y,j_z} + h.c., \qquad (3.44)$$

where

$$\epsilon(k_x, k_y) = (A \sin k_x \alpha_x + A \sin k_y \alpha_y) + \left(\Delta - 2B - 4B \sum_{i=x,y} \sin^2 \frac{k_i a}{2}\right)\beta.$$
$$(3.45)$$

Here, c_{k_x,k_y,j_z} is a four-component spinor. One can find the surface state solution by means of exact diagonalization.

3.6 Parity at the Time Reversal Invariant Momenta

We have constructed a lattice model by mapping the continuous model onto a lattice model. In the continuous model, the energy gap of the conduction and valence band opens near $k = 0$. In the mapping, k is replaced by $\frac{1}{a}\sin ka$. Since $\sin ka$ has two zero points at $ka = 0$ and $ka = \pi$, this property may make the two models topologically distinct. The topology of a system should be determined by the band structure of the whole Brillouin zone, not simply by the asymptotic behavior near a single point. In this section, we calculate the parity of the eigenstates at time reversal invariant momenta, which may reveal whether the lattice model is topologically trivial or nontrivial. We shall find that the parity of the eigenstates will change when the energy gap between the two bands closes and reopens, which accompanies a topological quantum phase transition. Readers can come back to this section after reading Chap. 4.

The parity operation π changes a right-handed system into a left-handed system:

$$\pi^\dagger \mathbf{x} \pi = -\mathbf{x}, \tag{3.46a}$$

$$\pi^\dagger \mathbf{p} \pi = -\mathbf{p}. \tag{3.46b}$$

π is not only unitary but also Hermitian:

$$\pi^\dagger = \pi^{-1} = \pi \tag{3.47}$$

and $\pi^2 = 1$. Hence, its eigenvalue is either $+1$ or -1. For a system with a parity symmetry, the energy eigenstates must be symmetric $(+1)$ or antisymmetric (-1):

$$\pi\phi(\mathbf{x}) = \phi(-\mathbf{x}) = \pm\phi(\mathbf{x}) \tag{3.48}$$

if they are nondegenerate. In the Dirac equation, the full parity operator P needs to be augmented with a unitary operator β [3]:

$$P = \pi\beta, \tag{3.49}$$

such that

$$P\alpha_i P = -\alpha_i, \, P\beta P = \beta. \tag{3.50}$$

In this way, the Dirac equation is invariant under parity P.

3.6.1 One-Dimensional Lattice Model

We begin with the one-dimensional lattice model,

$$H = A \sin k_x a \alpha_x + \left(\Delta - 4B \sin^2 \frac{k_x a}{2} \right) \beta. \tag{3.51}$$

The eigenvalues are doubly degenerate:

$$E_\pm = \pm \sqrt{A^2 \sin^2 k_x a + \left(\Delta - 4B \sin^2 \frac{k_x a}{2} \right)^2}. \tag{3.52}$$

Suppose the Fermi energy is zero. Then two occupied states have the negative energy and are time reversal partners with each other:

$$\psi_1 = \begin{pmatrix} -\frac{A \sin k_x a}{\sqrt{2E_+(E_+ + \Delta - 4B \sin^2 \frac{k_x a}{2})}} \\ 0 \\ 0 \\ \frac{\Delta - 4B \sin^2 \frac{k_x a}{2} + E_+}{\sqrt{2E_+(E_+ + \Delta - 4B \sin^2 \frac{k_x a}{2})}} \end{pmatrix} \tag{3.53}$$

and

$$\psi_2 = \Theta \psi_1. \tag{3.54}$$

where Θ is the time reversal operator. (See Appendix B.2).

The system is invariant under parity P since

$$PH(k) = H(-k)P. \tag{3.55}$$

Note that k is now a good quantum number, not an operator. From this relation, two time reversal invariant momenta can be defined:

$$PH(\Gamma_i) = H(\Gamma_i)P. \tag{3.56}$$

In one dimension, one is $\Gamma_1 = 0$:

$$PH(\Gamma_1 = 0) = H(-\Gamma_1)P, \tag{3.57}$$

and the other is $\Gamma_2 = \frac{1}{2}K = \frac{\pi}{a}$ (K is the reciprocal lattice vector):

$$PH(\Gamma_2) = H(-\Gamma_2 + K)P. \tag{3.58}$$

We calculate the eigenvalue of the parity of the state $|\psi_1\rangle$:

$$\delta|_{k=\Gamma_i} = \langle \psi_1 | P | \psi_1 \rangle = \mathrm{sgn} \left(-\Delta + 4B \sin^2 \frac{\Gamma_i a}{2} \right). \tag{3.59}$$

At the two time reversal invariant points, we have

$$\delta|_{ka=0} = \mathrm{sgn}(-\Delta), \tag{3.60a}$$

$$\delta|_{ka=\pi} = \mathrm{sgn}(-\Delta + 4B). \tag{3.60b}$$

We notice that the parity changes sign at the points of $\Delta = 0$ and $\Delta = 4B$, where the energy gap closes. The Z_2 index ν is determined by

$$(-1)^{\nu} = \delta|_{ka=0}\delta|_{ka=\pi} = \text{sgn}(\Delta)\text{sgn}(\Delta - 4B). \tag{3.61}$$

Thus, there are two distinct values of $(-1)^{\nu}$, $+1$ or -1. Correspondingly, $\nu = 0$ or 1. Therefore, for $0 < \Delta^2 < 4\Delta B$, the Z_2 index

$$\nu = 1 \tag{3.62}$$

which shows that the system is topologically nontrivial.

3.6.2 Two-Dimensional Lattice Model

For a two-dimensional lattice model,

$$H = A \sum_{i=x,y} \sin k_i a \alpha_i + \left(\Delta - 4B \sum_{i=x,y} \sin^2 \frac{k_i a}{2}\right)\beta. \tag{3.63}$$

The two energy eigenstates with the negative energy are

$$\psi_1 = \begin{pmatrix} \frac{-A(\sin k_x a - i \sin k_y a)}{\sqrt{2E_+(E_+ + \Delta - 4B(\sin^2 \frac{k_x a}{2} + \sin^2 \frac{k_y a}{2}))}} \\ 0 \\ 0 \\ \frac{\Delta - 4B(\sin^2 \frac{k_x a}{2} + \sin^2 \frac{k_y a}{2}) + E_+}{\sqrt{2E_+(E_+ + \Delta - 4B(\sin^2 \frac{k_x a}{2} + \sin^2 \frac{k_y a}{2}))}} \end{pmatrix} \tag{3.64}$$

and

$$\psi_2 = \Theta \psi_1. \tag{3.65}$$

The corresponding energy eigenvalue is

$$E_- = -\sqrt{A^2 \sum_{i=x,y} \sin^2 k_i a + \left(\Delta - 4B \sum_{i=x,y} \sin^2 \frac{k_i a}{2}\right)^2}. \tag{3.66}$$

The parity or the δ quantity at the time reversal invariant momenta is

$$\delta|_{k=\Gamma_i} = \langle \psi_1 | P | \psi_1 \rangle = \text{sgn}\left(-\Delta + 4B \sum_{i=x,y} \sin^2 \frac{\Gamma_i a}{2}\right). \tag{3.67}$$

In two dimensions, there are the four time reversal invariant momenta, $\Gamma_i a = (0,0)$, $\Gamma_i a = (0,\pi)$, $\Gamma_i a = (\pi,0)$, and $\Gamma_i a = (\pi,\pi)$. At these points, the parity of the state ψ_1 is

$$\delta|_{\Gamma_i a=(0,0)} = -\text{sgn}(\Delta), \tag{3.68a}$$

$$\delta|_{\Gamma_i a=(0,\pi)} = \text{sgn}(-\Delta + 4B), \tag{3.68b}$$

$$\delta|_{\Gamma_i a=(\pi,0)} = \text{sgn}(-\Delta + 4B), \tag{3.68c}$$

$$\delta|_{\Gamma_i a=(\pi,\pi)} = \text{sgn}(-\Delta + 8B). \tag{3.68d}$$

As a result,

$$(-1)^\nu = \text{sgn}(\Delta)\left[\text{sgn}(-\Delta + 4B)\right]^2 \text{sgn}(\Delta - 8B). \tag{3.69}$$

Therefore, we have a nontrivial index

$$\nu = 1 \tag{3.70}$$

for $0 < \Delta^2 < 8\Delta B$.

However, it is noted that $\delta|_{ka=(0,\pi)} = \delta|_{ka=(\pi,0)}$ discontinues at $\Delta = 4B$. Although the index is equal to 1 near the point, there exists a topological quantum phase transition. Both phases are topologically nontrivial. Accompanying the transition, the spin current around the boundary will change its sign.

3.6.3 Three-Dimensional Lattice Model

For a three-dimensional lattice model,

$$H = A \sum_{\alpha=x,y,z} \sin k_\alpha a \alpha_\alpha + \left(\Delta - 4B \sum_{\alpha=x,y,z} \sin^2 \frac{k_\alpha a}{2}\right)\beta. \tag{3.71}$$

The two energy eigenstates are

$$\psi_1 = \begin{pmatrix} \frac{-A(\sin k_x a - i \sin k_y a)}{\sqrt{2E_+(E_+ + \Delta - 4B\sum_{\alpha=x,y,z}\sin^2\frac{k_\alpha a}{2})}} \\ \frac{A\sin k_z}{\sqrt{2E_+(E_+ + \Delta - 4B\sum_{\alpha=x,y,z}\sin^2\frac{k_\alpha a}{2})}} \\ 0 \\ \frac{\Delta - 4B\sum_{\alpha=x,y,z}\sin^2\frac{k_\alpha a}{2} + E_+}{\sqrt{2E_+(E_+ + \Delta - 4B\sum_{\alpha=x,y,z}\sin^2\frac{k_\alpha a}{2})}} \end{pmatrix} \tag{3.72}$$

and

$$\psi_2 = \Theta\psi_1. \tag{3.73}$$

The corresponding negative energy is

$$E_- = -\sqrt{A^2 \sum_{\alpha=x,y,z} \sin^2 k_\alpha a + \left(\Delta - 4B \sum_{\alpha=x,y,z} \sin^2 \frac{k_\alpha a}{2}\right)^2}. \tag{3.74}$$

The parity at the time reversal invariant momenta is

$$\delta|_{k=\Gamma_i} = \langle \psi_1 | P | \psi_1 \rangle = \mathrm{sgn}\left(-\Delta + \sum_{\alpha=x,y,z} 4B \sin^2 \frac{\Gamma_\alpha a}{2}\right). \tag{3.75}$$

At the eight time reversal invariant points,

$$\delta|_{\Gamma_i a=(0,0,0)} = -\mathrm{sgn}(\Delta), \tag{3.76a}$$

$$\delta|_{\Gamma_i a=(0,0,\pi)} = \delta|_{\Gamma_i a=(0,\pi,0)} = \delta|_{\Gamma_i a=(\pi,0,0)} = \mathrm{sgn}(-\Delta + 4B), \tag{3.76b}$$

$$\delta|_{\Gamma_i a=(0,\pi,\pi)} = \delta|_{\Gamma_i a=(\pi,\pi,0)} = \delta|_{\Gamma_i a=(\pi,0,\pi)} = \mathrm{sgn}(-\Delta + 8B), \tag{3.76c}$$

$$\delta|_{\Gamma_i a=(\pi,\pi,\pi)} = \mathrm{sgn}(-\Delta + 12B). \tag{3.76d}$$

For $k_x = 0$,

$$(-1)^{\nu_1} = \delta|_{\Gamma_i a=(0,0,0)} \delta|_{\Gamma_i a=(0,0,\pi)} \delta|_{\Gamma_i a=(0,\pi,0)} \delta|_{\Gamma_i a=(0,\pi,\pi)}$$
$$= \mathrm{sgn}(\Delta)\mathrm{sgn}(\Delta - 8B). \tag{3.77}$$

For $k_y = 0$,

$$(-1)^{\nu_2} = \delta|_{\Gamma_i a=(0,0,0)} \delta|_{\Gamma_i a=(0,0,\pi)} \delta|_{\Gamma_i a=(\pi,0,0)} \delta|_{\Gamma_i a=(\pi,0,\pi)}$$
$$= \mathrm{sgn}(\Delta)\mathrm{sgn}(\Delta - 8B). \tag{3.78}$$

For $k_z = 0$,

$$(-1)^{\nu_3} = \delta|_{\Gamma_i a=(0,0,0)} \delta|_{\Gamma_i a=(0,\pi,0)} \delta|_{\Gamma_i a=(\pi,0,0)} \delta|_{\Gamma_i a=(\pi,\pi,0)}$$
$$= \mathrm{sgn}(\Delta)\mathrm{sgn}(\Delta - 8B). \tag{3.79}$$

For $k_x a = \pi$,

$$(-1)^{\nu_1'} = \delta|_{\Gamma_i a=(\pi,0,0)} \delta|_{\Gamma_i a=(\pi,0,\pi)} \delta|_{\Gamma_i a=(\pi,\pi,0)} \delta|_{\Gamma_i a=(\pi,\pi,\pi)}$$
$$= \mathrm{sgn}(\Delta - 4B)\mathrm{sgn}(\Delta - 12B). \tag{3.80}$$

The prime index ν_0 is determined by the product of parities at the eight time reversal invariant points.

$$(-1)^{\nu_0} = \prod_i \delta_i = (-1)^{\nu_1 + \nu_1'}$$

$$= \text{sgn}(\Delta)\text{sgn}(\Delta - 4B)\text{sgn}(\Delta - 8B)\text{sgn}(\Delta - 12B). \qquad (3.81)$$

Thus, for $B > 0$,

$$(\nu_0; \nu_1, \nu_2, \nu_3) = (0; 0, 0, 0), \text{ for } \Delta < 0, \qquad (3.82)$$

$$(\nu_0; \nu_1, \nu_2, \nu_3) = (1; 1, 1, 1), \text{ for } 0 < \Delta < 4B, \qquad (3.83)$$

$$(\nu_0; \nu_1, \nu_2, \nu_3) = (0; 1, 1, 1), \text{ for } 4B < \Delta < 8B, \qquad (3.84)$$

$$(\nu_0; \nu_1, \nu_2, \nu_3) = (1; 0, 0, 0), \text{ for } 8B < \Delta < 12B, \qquad (3.85)$$

$$(\nu_0; \nu_1, \nu_2, \nu_3) = (0; 0, 0, 0), \text{ for } \Delta > 12B. \qquad (3.86)$$

The system is topologically nontrivial only if $0 < \Delta < 4B$ and $8B < \Delta < 12B$.

3.7 Summary

In summary, a minimal lattice model for topological insulator is established in one, two, and three dimensions. According to the parity of the eigenstates at the time reversal invariant momenta, we conclude that (suppose B positive)

1. In one dimension, it is topologically nontrivial for $0 < \Delta < 4B$.
2. In two dimensions, it is topologically nontrivial for $0 < \Delta < 4B$ and $4B < \Delta < 8B$.
3. In three dimensions, it is topologically nontrivial for $0 < \Delta < 4B$ and $8B < \Delta < 12B$.

References

1. S.Q. Shen, W.Y. Shan, H.Z. Lu, SPIN **1**, 33 (2011)
2. K. Imura, A. Yamakage, S.J. Mao, A. Hotta, Y. Kuramoto, Phys. Rev. B **82**, 085118 (2010)
3. J.D. Bjorken, S.D. Drell, *Relativistic Quantum Mechanics* (McGraw-Hill, New York, 1964)

Chapter 4
Topological Invariants

Abstract There are two classes of topological invariants for topological insulators. One is characterized by the elements of the group Z, which consists of all integers. For example, the integer quantum Hall effect is characterized by an integer n, that is, the filling factor of electrons. The other is by the elements of the group Z_2, which consists of 0 and 1 or 1 and −1. In a topological insulator with time reversal symmetry, 0 and 1 represent the existence of odd and even numbers of the surface states in three dimensions or the even and odd number pairs of helical edge states in two dimensions, respectively.

Keywords Bloch theorem • Berry phase • Charge pump • Spin pump • Laughlin argument • Chern number • The Z_2 index

4.1 Bloch Theorem and Band Theory

A Bloch wave or a Bloch state, named after Felix Bloch, is the wave function of an electron in a periodic potential. Let us consider a Hamiltonian $\mathcal{H}(\mathbf{r}) = \mathcal{H}(\mathbf{r} + \mathbf{R})$ in a periodic potential. Bloch's theorem states that the eigenfunction for such a system must be of the form

$$|\psi_{n,\mathbf{k}}(\mathbf{r})\rangle = e^{i\mathbf{k}\cdot\mathbf{r}}|u_{n,\mathbf{k}}(\mathbf{r})\rangle, \tag{4.1}$$

where $u_{n,\mathbf{k}}(\mathbf{r})$ has the period of the crystal lattice $\mathbf{R}$ with $u_{n,\mathbf{k}}(\mathbf{r}) = u_{n,\mathbf{k}}(\mathbf{r} + \mathbf{R})$. $u_{n,\mathbf{k}}(\mathbf{r})$ is the cell periodic eigenstate of $H(\mathbf{k}) = e^{-i\mathbf{k}\cdot\mathbf{r}}\mathcal{H}(\mathbf{r})e^{i\mathbf{k}\cdot\mathbf{r}}$,

$$H(\mathbf{k})|u_{n,\mathbf{k}}(\mathbf{r})\rangle = E_{n,\mathbf{k}}|u_{n,\mathbf{k}}(\mathbf{r})\rangle. \tag{4.2}$$

The corresponding energy eigenvalues satisfy, $E_n(\mathbf{k}) = E_n(\mathbf{k} + \mathbf{K})$, periodic with periodicity $\mathbf{K}$ of a reciprocal lattice vector. The energies associated with the index n vary continuously with the wave vector $\mathbf{k}$ and form an energy band identified by

S.-Q. Shen, *Topological Insulators: Dirac Equation in Condensed Matters*,
Springer Series in Solid-State Sciences 174, DOI 10.1007/978-3-642-32858-9_4,
© Springer-Verlag Berlin Heidelberg 2012

the band index n. The eigenvalues for given n are periodic in $\mathbf{k}$; all distinct values of $E_n(\mathbf{k})$ are located within the first Brillouin zone of the reciprocal lattice. See Ref. [1].

According to the Pauli exclusion principle, each state can be occupied at most by one electron. Electrons will fill lower energy states first and consequently form the Fermi sea for a finite density of electrons. The highest energy of the occupied states is called the Fermi level or Fermi energy. Near the Fermi level, if the band is partially occupied, it is a metallic state. In this case, when an external field is applied to the system, the field will force electrons to shift away from the equilibrium position and gain a nonzero total momentum to form a flow of electric current. If the band is fully filled, and there exists an energy gap between the filled or valence band and the unfilled or conduction band, it is an insulating state. In this case, a weak external field cannot force the electrons to move away from the occupied states to circulate a flow of electric current. This is the picture for the band insulator. The size of the energy gap serves as a dividing line between semiconductors and insulators. If the energy gap is smaller than 4 eV (roughly), the electrons can be excited easily from the valence band to conduction band at finite temperatures, although the fully filled band does not contribute to electrical conductivity at absolute zero temperature. Thus, a semiconductor has a smaller energy gap than an insulator.

4.2 Berry Phase

The choice of $|u_{n,\mathbf{k}}\rangle$ in Eq. (4.2) is not unique. For example, there is always a $U(1)$, that is, a phase uncertainty,

$$|u_{n,\mathbf{k}}\rangle \to e^{\mathrm{i} f(\mathbf{k})} |u_{n,\mathbf{k}}\rangle, \tag{4.3}$$

keeping Eq. (4.2) invariant. A definite set of phase choice in the Brillouin zone is called a definite gauge [2]. For a time reversal invariant system, there always exists a continuous gauge throughout the Brillouin zone. For a time reversal broken system with nonzero Chern number, there is no such a gauge that continuous gauges have to be defined in different patches of the Brillouin zone [2, 3]. However, any physical observable must be gauge independent.

Consider the system Hamiltonian that varies with time through a parameter $\mathbf{R} \to \mathbf{R}(t)$. We are interested in a cyclic evolution of the system from $t = 0$ to T such that $\mathbf{R}(t = 0) = \mathbf{R}(t = T)$. The parameter $\mathbf{R}(t)$ changes very slowly along a closed path C in the parameter space. To solve the problem, we first introduce an instantaneous orthogonal basis from the instantaneous eigenstates of $H(\mathbf{R}(t))$ at time t or each value of $\mathbf{R}(t)$:

$$H(\mathbf{R}(t)) |u_n(\mathbf{R}(t))\rangle = \varepsilon_n(\mathbf{R}(t)) |u_n(\mathbf{R}(t))\rangle. \tag{4.4}$$

This equation does not completely determine the basis function of $|u_n(\mathbf{R}(t))\rangle$ due to the phase uncertainty. However, we can require that the functions are smooth and single valued along the closed path. The equation also does not describe correctly the time evolution of the quantum states. Instead the quantum state should be governed by the time-dependent Schrödinger equation,

$$i\hbar\partial_t |\Phi(t)\rangle = H(\mathbf{R}(t)) |\Phi(t)\rangle . \tag{4.5}$$

In the adiabatic approximation [4], the system will stay at one of the instantaneous eigenstates (usually we choose the lowest energy state or the ground state) if the instantaneous state is well separated with the others and the time evolution is very slowly. In this case, this wave function can be related to $|u_n(\mathbf{R}(t))\rangle$:

$$|\Phi(t)\rangle = e^{i\gamma_c(t)} \exp\left[-\frac{i}{\hbar} \int_0^t dt' \varepsilon_n(\mathbf{R}(t'))\right] |u_n(\mathbf{R}(t))\rangle \tag{4.6}$$

and

$$\partial_t \gamma_c(t) = i \langle u_n(t)| \partial_t |u_n(t)\rangle . \tag{4.7}$$

In the parameter space, the phase factor can be expressed as a path integral

$$\gamma_c = \int_C d\mathbf{R} \cdot \mathbf{A}_n(\mathbf{R}), \tag{4.8}$$

where $\mathbf{A}_n(\mathbf{R})$ is a vector

$$\mathbf{A}_n(\mathbf{R}) = i \langle u_n(\mathbf{R}(t))| \nabla_R |u_n(\mathbf{R}(t))\rangle . \tag{4.9}$$

This vector is called the Berry connection or the Berry vector potential. In addition to the dynamic phase which is determined by integrating over $\varepsilon_n(\mathbf{R}(t'))$, the state $|\Phi(t)\rangle$ will acquire an additional phase γ_c during the adiabatic evolution.

Since $\mathbf{A}_n(\mathbf{R})$ is gauge dependent, it becomes

$$\mathbf{A}_n(\mathbf{R}) \rightarrow \mathbf{A}_n(\mathbf{R}) - \nabla_R \chi \tag{4.10}$$

if we make a gauge transformation

$$|u_n(\mathbf{R}(t))\rangle \rightarrow e^{i\chi(\mathbf{R})} |u_n(\mathbf{R}(t))\rangle . \tag{4.11}$$

Thus, the phase γ_c will be changed by $\chi(\mathbf{R}(t = T)) - \chi(\mathbf{R}(t = 0))$ for the initial and final points. For a cyclic evolution of the system along a closed path C with $\mathbf{R}(0) = \mathbf{R}(T)$, the single-valued condition of the wave function requires

$$\chi(\mathbf{R}(T)) - \chi(\mathbf{R}(0)) = 2m\pi \tag{4.12}$$

with an integer m. Therefore, for a closed path C, γ_c is independent of the gauge and now is known as the Berry phase:

$$\gamma_c = \oint_C d\mathbf{R} \cdot \mathbf{A}_n(\mathbf{R}). \tag{4.13}$$

By using the Stokes' theorem, γ_c can be expressed as an area integral

$$\gamma_c = \int_S d\mathbf{S} \cdot \Omega(\mathbf{R}), \tag{4.14}$$

where the Berry curvature from the Berry connection is defined as

$$\Omega^n(\mathbf{R}) = \nabla_{\mathbf{R}} \times \mathbf{A}_n(\mathbf{R}). \tag{4.15}$$

Its components are

$$
\begin{aligned}
\Omega^n_{\mu\nu}(\mathbf{R}) &= \partial_\mu (\mathbf{A}_n)_\nu - \partial_\nu (\mathbf{A}_n)_\mu \\
&= i \left(\langle \partial_\mu u_n(\mathbf{R}) | \partial_\nu u_n(\mathbf{R}) \rangle - \langle \partial_\nu u_n(\mathbf{R}) | \partial_\mu u_n(\mathbf{R}) \rangle \right),
\end{aligned} \tag{4.16}
$$

where we denote $\partial/\partial R_\mu$ by ∂_μ. The Berry curvature Ω is analogous to the magnetic field in electrodynamics. Using the completeness relation for the basis,

$$\sum_n |u_n(\mathbf{R})\rangle \langle u_n(\mathbf{R})| = 1 \tag{4.17}$$

and the identity

$$\langle u_m(\mathbf{R})| \nabla_{\mathbf{R}} |u_n(\mathbf{R})\rangle = \frac{\langle u_m(\mathbf{R})| \nabla_{\mathbf{R}} H(\mathbf{R}) |u_n(\mathbf{R})\rangle}{E_n - E_m}$$

($m \neq n$), the Berry curvature has an alternative expression

$$\Omega^n = \mathrm{Im} \sum_{m \neq n} \frac{\langle u_n(\mathbf{R})| \nabla_{\mathbf{R}} H(\mathbf{R}) |u_m(\mathbf{R})\rangle \times \langle u_m(\mathbf{R})| \nabla_{\mathbf{R}} H(\mathbf{R}) |u_n(\mathbf{R})\rangle}{(E_n - E_m)^2}.$$

Consider a two-level system as an example. The general Hamiltonian describing a two-level problem has the form

$$H = \frac{1}{2} \begin{pmatrix} Z & X - iY \\ X + iY & -Z \end{pmatrix} = \frac{1}{2}\mathbf{R} \cdot \sigma. \tag{4.18}$$

The energy eigenvalues are $E_\pm = \pm R = \pm\sqrt{X^2 + Y^2 + Z^2}$ and the two levels cross at the degeneracy point of $R = 0$. The gradient of the Hamiltonian is $\nabla_R H = \frac{1}{2}\sigma$, and we find that the Berry curvature has its vector form

$$\Omega = \frac{1}{2}\frac{\mathbf{R}}{R^2}.$$ (4.19)

This curvature can be regarded as a field generated by a magnetic monopole at the origin $\mathbf{R} = 0$. Integrating the Berry curvature over a sphere containing the monopole, we have

$$\frac{1}{2\pi}\int_S d\mathbf{S}\cdot\Omega = 1.$$ (4.20)

The divergence of Ω has the property

$$\nabla_{\mathbf{R}} \cdot \Omega = 2\pi\delta(\mathbf{R}).$$ (4.21)

Thus, a point-like "magnetic monopole" is located at $R = 0$, which generates the Berry curvature.

In the Bloch bands, the Berry curvature is defined as

$$\Omega^n(\mathbf{k}) = \nabla_{\mathbf{k}} \times \langle u_n(\mathbf{k})| \, i \, \nabla_{\mathbf{k}} |u_n(\mathbf{k})\rangle.$$ (4.22)

Since the two points $\mathbf{k}$ and $\mathbf{k} + \mathbf{K}$ in the Brillouin zone can be identified as the same point, where $\mathbf{K}$ is the reciprocal lattice vector, a closed path can be realized when $\mathbf{k}$ sweeps the whole Brillouin zone. In this case, the Berry phase across the Brillouin zone becomes [5]

$$\gamma_c = \int_{BZ} d\mathbf{k} \cdot \langle u_n(\mathbf{k})| \, i \, \nabla_{\mathbf{k}} |u_n(\mathbf{k})\rangle.$$ (4.23)

4.3 Quantum Hall Conductance and Chern Number

The Hall conductance in a two-dimensional band insulator can be expressed in terms of the Berry curvature:

$$\sigma_{xy} = \frac{e^2}{\hbar} \int_{BZ} \frac{d\mathbf{k}}{(2\pi)^2}\Omega_{k_x,k_y} = n\frac{e^2}{h},$$ (4.24)

which is quantized for an integer n (including zero). Consider a crystal under the perturbation of a weak electric field $\mathbf{E}$. Usually the electrostatic potential $\phi(r)$ which produces an electric field $\mathbf{E} = -\nabla\phi$ varies linearly in space and breaks the translational symmetry. If the electric field enters the Hamiltonian through the electrostatic potential $\phi(r)$, the wave vector is no longer a good quantum number and the Bloch theorem fails to apply to the problem. To avoid this difficulty, one can introduce a uniform vector potential $\mathbf{A}(t)$ that changes in time such that $\partial_t \mathbf{A}(t) = -\mathbf{E}$. The Hamiltonian is written as

$$H(t) = \frac{1}{2m}(\mathbf{p} + e\mathbf{A}(t))^2 + V(r).$$ (4.25)

Here we take the elementary charge of electron $-e$ ($e > 0$). Thus, in this way, the lattice translational symmetry is preserved, and the momentum $\mathbf{p}$ is still a good quantum number. In the momentum space, $\mathbf{p} = \hbar \mathbf{q}$, we have

$$H(\mathbf{q}, t) = H\left(\mathbf{q} + \frac{e}{\hbar}\mathbf{A}(t)\right). \tag{4.26}$$

Now we introduce the gauge-invariant crystal momentum

$$\mathbf{k} = \mathbf{q} + \frac{e}{\hbar}\mathbf{A}(t). \tag{4.27}$$

Since $\mathbf{q}$ is a good quantum number, that is, $d\mathbf{q}/dt = 0$, it follows that

$$\frac{d\mathbf{k}}{dt} = -\frac{e}{\hbar}\mathbf{E}. \tag{4.28}$$

The velocity operator is defined by

$$\mathbf{v} = \frac{d\mathbf{r}}{dt} = \frac{i}{\hbar}[H, \mathbf{r}]. \tag{4.29}$$

In the momentum space, it becomes

$$\mathbf{v}(\mathbf{q}) = \mathrm{e}^{-i\mathbf{q}\cdot\mathbf{r}}\frac{i}{\hbar}[H, \mathbf{r}]\mathrm{e}^{i\mathbf{q}\cdot\mathbf{r}} = \frac{1}{\hbar}\nabla_{\mathbf{q}}H(\mathbf{q}, t). \tag{4.30}$$

The presence of $\mathbf{A}(t)$ makes the problem time dependent. The wave function for the quantum state $\psi(t)$ is governed by the time-dependent Schrödinger equation,

$$i\hbar\partial_t|\psi(t)\rangle = H(t)|\psi(t)\rangle. \tag{4.31}$$

Using the instantaneous eigenstates as the basis, we can expand the wave function $\psi(t)$ in terms of the instantaneous eigenstates $|u_n(t)\rangle$ and eigenvalues $E_n(t)$:

$$|\psi(t)\rangle = \sum_n \exp\left(\frac{1}{i\hbar}\int_{t_0}^t dt' E_n(t')\right) a_n(t)|u_n(\mathbf{q}, t)\rangle. \tag{4.32}$$

Then the Schrödinger equation is reduced to

$$\frac{da_n(t)}{dt} = -\sum_m a_m(t)\langle u_n(t)|\partial_t u_m(t)\rangle \exp\left(-i\int_{t_0}^t dt'\omega_{mn}(t')\right). \tag{4.33}$$

where $\omega_{mn}(t) = (E_m(t') - E_n(t'))/\hbar$. For our purpose, we consider an adiabatic process that the vector parameter $\mathbf{R}(t)$ varies with time very slowly, and

$$\langle u_n(\mathbf{q}, t)|\partial_t u_n(\mathbf{q}, t)\rangle = \partial_t \mathbf{R} \cdot \langle u_n(\mathbf{q}, \mathbf{R})| \nabla_{\mathbf{R}} |u_n(\mathbf{q}, \mathbf{R})\rangle \ll 1. \qquad (4.34)$$

In the limit of $\partial_t \mathbf{R} = 0$, we have

$$\partial_t a_n = 0. \qquad (4.35)$$

If the system is initially in the eigenstate $|u_n(\mathbf{q}, t = 0)\rangle$, it will stay in that state afterward. This is the quantum adiabatic theorem [4].

Now we consider the case that $\partial_t \mathbf{R} \neq 0$ but still very small. Suppose the initial state has $a_n(0) = 1$ and $a_m(0) = 0$ for all $m \neq n$. We apply the time-dependent perturbation theory to calculate the quantum correction to the states due to the perturbation of the electric field. The zero-order perturbation gives that $a_m^{(0)} = \delta_{m,n}$. Thus, the first-order perturbation $a_m^{(1)}$ is given by

$$\frac{da_m^{(1)}(t)}{dt} = -\langle u_m(\mathbf{q}, t)|\partial_t u_n(\mathbf{q}, t)\rangle \exp\left(-i \int_{t_0}^t dt' \omega_{nm}(t')\right). \qquad (4.36)$$

For $m = n$, $\frac{da_m^{(1)}(t)}{dt} = 0$. Thus, we have

$$a_n^{(1)}(t) = 0. \qquad (4.37)$$

For $m \neq n$,

$$a_m^{(1)}(t) = -i\hbar \frac{\langle u_m(\mathbf{q}, t)|\partial_t u_n(\mathbf{q}, t)\rangle}{E_n - E_m} \exp\left(-i \int_{t_0}^t dt' \omega_{nm}(t')\right). \qquad (4.38)$$

Thus, the wave function up to the first-order perturbation is given by

$$|u_n(t)\rangle \rightarrow |u_n(\mathbf{q}, t)\rangle - i\hbar \sum_{m \neq n} |u_m(\mathbf{q}, t)\rangle \frac{\langle u_m(\mathbf{q}, t)|\partial_t u_n(\mathbf{q}, t)\rangle}{E_n - E_m}. \qquad (4.39)$$

Using the velocity operator in Eq. (4.30), the average velocity in the state after the perturbation becomes

$$\mathbf{v}_n(\mathbf{q}) = -i \sum_{m \neq n} \left(\frac{\langle u_n(\mathbf{q}, t)| \nabla_{\mathbf{q}} H |u_m(\mathbf{q}, t)\rangle \langle u_m(\mathbf{q}, t)|\partial_t u_n(\mathbf{q}, t)\rangle}{E_n - E_m} - h.c \right)$$

$$+ \frac{1}{\hbar} \nabla_{\mathbf{q}} E_n(\mathbf{q}). \qquad (4.40)$$

Furthermore, using the identity

$$\langle u_n(\mathbf{q}, t)| \nabla_{\mathbf{q}} H |u_m(\mathbf{q}, t)\rangle = (E_n - E_m) \langle \nabla_{\mathbf{q}} u_n(\mathbf{q}, t)|u_m(\mathbf{q}, t)\rangle, \qquad (4.41)$$

the expression can be simplified in a compact form:

$$\mathbf{v}_n(\mathbf{q}) = \frac{1}{\hbar}\nabla_{\mathbf{q}}E_n(\mathbf{q}) - \Omega^n_{\mathbf{q},t},$$ (4.42)

where

$$\Omega^n_{\mathbf{q},t} = i\left(\langle\nabla_{\mathbf{q}}u_n|\partial_t u_n\rangle - \langle\partial_t u_n|\nabla_{\mathbf{q}}u_n\rangle\right).$$ (4.43)

Thus, in the presence of an electric field, an electron can acquire an anomalous transverse velocity proportional to the Berry curvature of the energy band [6, 7].

It follows from Eqs. (4.27) and (4.28) that

$$\nabla_{\mathbf{q}} = \nabla_{\mathbf{k}},$$ (4.44a)

$$\partial_t = \partial_t\mathbf{k}\cdot\nabla_{\mathbf{k}} = -\frac{e}{\hbar}\mathbf{E}\cdot\nabla_{\mathbf{k}}.$$ (4.44b)

Thus, the velocity is reduced to

$$\mathbf{v}_n(\mathbf{q}) = \frac{1}{\hbar}\nabla_{\mathbf{k}}E_n(\mathbf{k}) - \frac{e}{\hbar}\mathbf{E}\times\Omega^n(\mathbf{k})$$ (4.45)

where

$$\Omega^n(\mathbf{k}) = \nabla_{\mathbf{k}}\times\langle u_n(\mathbf{k})|\,i\,\nabla_{\mathbf{k}}\,|u_n(\mathbf{k})\rangle$$
$$= i\,\langle\nabla_{\mathbf{k}}u_n(\mathbf{k})|\times|\nabla_{\mathbf{k}}u_n(\mathbf{k})\rangle.$$ (4.46)

Thus, the external field produces a transverse velocity in an adiabatic process. The electric current in the presence of $\mathbf{E}$ is defined by

$$\mathbf{j} = -e\sum_n\int\frac{d\mathbf{k}}{(2\pi)^2}\mathbf{v}_n(\mathbf{k})f(k),$$ (4.47)

where $f(k)$ is the Fermi-Dirac distribution function. Suppose all bands below the Fermi level are fully filled. The sum over the first term in the velocity in Eq. (4.45) becomes zero, and the second term gives a Hall current

$$j_\alpha = \sigma_H\epsilon_{\alpha\beta}E_\beta$$ (4.48)

with

$$\sigma_H = \frac{e^2}{h}\frac{1}{2\pi}\sum_n\int_{BZ}d\mathbf{k}\Omega^n_{k_x,k_y}.$$ (4.49)

The integral runs over the first Brillouin zone, and

$$\Omega^n_{k_x,k_y} = \Omega^n_{k_x+\pi,k_y} = \Omega^n_{k_x,k_y+\pi}. \tag{4.50}$$

Hence, the first Brillouin zone forms a closed torus. In this expression, we assume that all bands are fully filled, and there exists an energy gap between the filled band or valence band and the unfilled band or conduction band. The integral over a closed torus gives an integer ν (including zero),

$$\sigma_H = \nu \frac{e^2}{h}. \tag{4.51}$$

This result can also be derived from the Kubo formula explicitly (see Appendix A.1).

4.4 Electric Polarization in a Cyclic Adiabatic Evolution

Electric polarization **P** is the electric dipole momentum per volume in dielectric media, which is one of the essential concepts in electrodynamics. It is an intensive vector quantity that carries the meaning of the dipole moment per unit volume. For example, in a ferroelectric material, the electric polarization can present spontaneously. In the Maxwell's equation for the displacement **D**,

$$\nabla \cdot \mathbf{D} = -\rho(t), \tag{4.52}$$

where $\mathbf{D} = \epsilon_0 \mathbf{E} + \mathbf{P}$. Here **E** is the electric field, **P** is the polarization density, and $\rho(t)$ is the charge density. Consider a solid in which there is no electric field. The continuity equation $\partial_t \rho = -\nabla \cdot \mathbf{j}$ leads to

$$\nabla \cdot (\partial_t \mathbf{P} - \mathbf{j}) = 0, \tag{4.53}$$

where **j** is the macroscopic current density. In an adiabatic evolution of a system, up to a divergence-free part, the change in the polarization density in a cyclic evolution is given by

$$\Delta P_\alpha = \int_0^T dt j_\alpha. \tag{4.54}$$

This equation is the basis for the modern theory of polarization. In the early 1990s, it was realized that the polarization difference has its topological meaning and is actually related to the Berry phase [2, 8].

In an adiabatic process, it follows from Eq. (4.42) that

$$\Delta P_\alpha = e \sum_n \int_0^T dt \int_{BZ} \frac{d\mathbf{q}}{(2\pi)^d} \Omega^n_{q_\alpha,t}, \tag{4.55}$$

which is determined by the Berry curvature $\Omega^n_{q_\alpha,t}$. The summation runs over all the occupied bands. For a general purpose, we suppose that the adiabatic transformation is parameterized by a scalar $\lambda(t)$; it follows that [9]

$$\Delta P_\alpha = e \sum_n \int_{\lambda(0)}^{\lambda(T)} d\lambda \int_{BZ} \frac{d\mathbf{q}}{(2\pi)^d} \Omega^n_{q_\alpha,\lambda}, \tag{4.56}$$

where

$$\Omega^n_{q_\alpha,\lambda} = \partial_{q_\alpha} \mathbf{A}^n_\lambda - \partial_\lambda \mathbf{A}^n_{q_\alpha}. \tag{4.57}$$

In the course of a cyclic evolution, $\lambda(T)$ and $\lambda(0)$ will represent the same state. Consider the periodicity of the $\mathbf{q}$ space. The $q_\alpha - \lambda$ plane forms a close torus. It should be pointed out that the polarization is determined up to an uncertainty quantum. Since the integral does not track the history of λ, there is no information on how many cycles λ has gone through. For each cycle, an integer number ν of electrons are transported across the sample [2]:

$$\Delta P_\alpha = e\nu a, \tag{4.58}$$

where a is the lattice constant. Here the integer ν appears as a topological invariant for the adiabatic transport.

From the Bloch function, we can define the Wannier function associated with the lattice vector:

$$|\mathbf{R}, n\rangle = \frac{1}{(2\pi)^d} \int d\mathbf{k} e^{-i\mathbf{k}\cdot(\mathbf{R}-\mathbf{r})} |u_{n,\mathbf{k}}\rangle . \tag{4.59}$$

King-Smith and Vanderbilt [9] showed that the polarization can be defined by the sum over all the bands of the center of charge of the Wannier state associated with $\mathbf{R} = 0$:

$$\mathbf{P} = -e \sum_n \langle \mathbf{R} = 0, n| \mathbf{r} |\mathbf{R} = 0, n\rangle = -\frac{e}{2\pi} \oint d\mathbf{k} \cdot \mathbf{A}(k), \tag{4.60}$$

where $\mathbf{A}(\mathbf{k}) = i \sum_n \langle u_{n,\mathbf{k}}| \nabla_\mathbf{k} |u_{n,\mathbf{k}}\rangle$. Here we have used the relation $\mathbf{r} = i \nabla_\mathbf{k}$.

4.5　Thouless Charge Pump

In a cyclic adiabatic evolution of a one-dimensional insulator,

$$H(k, t + T) = H(k, t), \tag{4.61}$$

the charge pumped across the insulator is always an integer, which is defined as a topological invariant, that is, the electric polarization

$$\Delta P = \frac{e}{2\pi} \oint [A(k, T) - A(k, 0)] \, dk = nea. \tag{4.62}$$

Here we present an example to illustrate the process of the charge pump. The Rice-Mele model was introduced in the study of solitons in the polyenes in 1980s, and later used to study ferroelectricity [10]:

$$H = +h_{st}(t) \sum_i (-1)^i c_i^\dagger c_i + \frac{1}{2} \sum_{i=1}^N \left[t_0 + \delta(t)(-1)^i \right] c_i^\dagger c_{i+1} + h.c., \tag{4.63}$$

where

$$(\delta(t), h_{st}(t)) = \left(\delta_0 \cos \frac{2\pi t}{T}, h_0 \sin \frac{2\pi t}{T} \right) \tag{4.64}$$

and N is an even number. This is a time-dependent model: $\delta(t)$ denotes the displacements of the ith and $(i + 1)$th electrons from their respective equilibrium position in a staggered or dimerized form, and $\pm h_{st}(t)$ are the staggered on-site potentials. Both $\delta(t)$ and $h_{st}(t)$ are periodic functions of time t with a period T.

We consider a system with an even number $2N$ of lattice sites and take a periodic boundary condition. Performing the Fourier transformation,

$$a_k = \frac{1}{\sqrt{N}} \sum_{j \in 2n} c_j e^{-ikj}, \tag{4.65a}$$

$$b_k = \frac{1}{\sqrt{N}} \sum_{j \in 2n+1} c_j e^{-ikj}, \tag{4.65b}$$

the Hamiltonian is reduced to

$$H = \sum_k (a_k^\dagger, b_k^\dagger) \mathbf{d}(k, t) \cdot \sigma \begin{pmatrix} a_k \\ b_k \end{pmatrix}, \tag{4.66}$$

where

$$d_x(k, t) = \frac{1}{2}(t_0 + \delta(t)) + \frac{1}{2}(t_0 - \delta(t)) \cos k, \tag{4.67a}$$

$$d_y(k, t) = -\frac{1}{2}(t_0 - \delta(t)) \sin k, \tag{4.67b}$$

$$d_z(k, t) = h_{st}(t). \tag{4.67c}$$

The instantaneous dispersions of the two bands at time t are

$$\varepsilon_\pm(k, t) = \pm |\mathbf{d}(k, t)| \tag{4.68}$$

$$= \pm \sqrt{h_0^2 \sin^2 \frac{2\pi t}{T} + \delta_0^2 \cos^2 \frac{2\pi t}{T} \sin^2 \frac{k}{2} + t_0^2 \cos^2 \frac{k}{2}}. \tag{4.69}$$

The degeneracy points are $h_0 = 0$, or $\delta_0 = 0$, and $t_0 = 0$. The energy gap between two bands is $\Delta E = \min(2|t_0|, 2|h_0|, 2|\delta_0|)$. So the adiabatic condition requires that $T \gg \hbar/\min(2|t_0|, 2|h_0|, 2|\delta_0|)$. If the low band is fully filled, the charge pump in a cyclic adiabatic evolution is associated with the Chern number of the ground state $\Delta P = n_c ea$:

$$
\begin{aligned}
n_c &= \int_0^T dt \int_{BZ} \frac{dk}{2\pi} \Omega_{k,t}^n, \\
&= -\frac{1}{4\pi} \int dk \int_0^T dt \frac{\mathbf{d}(k,t) \cdot [\partial_k \mathbf{d}(k,t) \times \partial_t \mathbf{d}(k,t)]}{|\mathbf{d}(k,t)|^3} \\
&= -\mathrm{sgn}(t_0 h_0 \delta_0)
\end{aligned}
\tag{4.70}
$$

because the $k - t$ plane forms a closed torus due the periodicity of T. We find that the Chern number is $+1$ or -1 once $t_0 h_0 \delta_0 \neq 0$. A topological quantum phase transition occurs at the points of $h_0 = 0$, or $\delta_0 = 0$, $t_0 = 0$, where the Chern number changes its sign whenever any one of the parameters changes its sign.

The charge pumping can be understood based on the picture of the end states in an open chain. The Rice-Mele model is reduced to the Su-Schrieffer-Heeger model when $\delta(t) \neq 0$ and $h_{st}(t) = 0$. The solution of the end state in this model can be found in Sect. 5.1. At $t = 0$, $(\delta(t), h_{st}(t)) = (+\delta_0, 0)$. The hopping amplitudes along the chain starting from site $i = 1$ are $t_0 - \delta_0, t_0 + \delta_0, t_0 - \delta_0, t_0 + \delta_0, \cdots$. Assume $t_0 > \delta_0 > 0$. In this case, there exist two end states of zero energy at two ends of the chain, respectively, which are degenerate at $h_0 = 0$. At the half filling that one particle occupies two sites averagely, we suppose that the right end state is occupied and the left end state is empty. With increasing time t, the on-site energy $h_{st}(t)$ lifts the end mode away from the zero energy to the valence band: one is pushed to the positive band and the other to the negative band. At $t = T/2$, $(\delta(t), h_{st}(t)) = (-\delta_0, 0)$. The hopping amplitudes become $t_0 + \delta_0, t_0 - \delta_0, t_0 + \delta_0, t_0 - \delta_0, \cdots$. In this case, the two end states disappear as they have already evolved into the bulk states. When t continuously increases, the end states appear again. However, the occupied end state becomes the left one, and the right end state becomes empty. At $t = T$, $(\delta(t), h_{st}(t)) = (+\delta_0, 0)$. The hopping amplitudes go back to the case of $t = 0$. The Hamiltonian returns to the original one at $t = 0$. Although the energy eigenstates remain unchanged, due to the double degeneracy of the ground state at half filling, the electron configuration has changed: the electron in the right end state at $t = 0$ has been transferred to the left end state at $t = T$. In this way, one electron has been pumped from the left to right side. The instantaneous spectra of the Rice-Mele model in Eq. (4.63) are plotted in Fig. 4.1.

4.6   Fu-Kane Spin Pump

Fu and Kane proposed an electronic model with spin $\frac{1}{2}$ for spin pump by generalizing the spinless Rice-Mele model [11]:

Fig. 4.1 The instantaneous energy spectra of the Rice-Mele model. The *solid line* stands for the end state near the *right* end, while the *dashed line* for the state at the *left* side. It illustrates the evolution of the end state from the one side to the other. Here we take $\delta_0 = 0.2t_0$ and $h_0 = 0.5t_0$

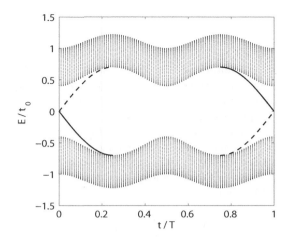

$$H = h_{st}(t) \sum_{i,\sigma=\pm1} (-1)^i c_{i,\sigma}^\dagger \sigma_{\sigma\sigma'}^z c_{i,\sigma'} + \frac{1}{2} \sum_{i,\sigma=\pm1} \left[t_0 + (-1)^i \delta(t) \right] c_{i,\sigma}^\dagger c_{i+1,\sigma} + h.c.,$$

(4.71)

where

$$(\delta(t), h_{st}(t)) = \left(\delta_0 \cos \frac{2\pi t}{T}, h_0 \sin \frac{2\pi t}{T} \right).$$

(4.72)

A magnetic staggered field is introduced to replace the on-site potential. We choose the eigenstates of σ_z as the basis, and set $\phi_{k,\uparrow}^\dagger = (a_{k,\uparrow}^\dagger, b_{k,\uparrow}^\dagger)$ and $\phi_{k,\downarrow}^\dagger = (a_{k,\downarrow}^\dagger, b_{k,\downarrow}^\dagger)$. The model is diagonalized in block with spin-up and spin-down:

$$H = \sum_k (\phi_{k,\uparrow}^\dagger, \phi_{k,\downarrow}^\dagger) \begin{pmatrix} d_+ \cdot \sigma & 0 \\ 0 & d_- \cdot \sigma \end{pmatrix} \begin{pmatrix} \phi_{k,\uparrow} \\ \phi_{k,\downarrow} \end{pmatrix},$$

(4.73)

where

$$(\mathbf{d}_\pm)_x = \frac{1}{2}(t_0 + \delta(t)) + \frac{1}{2}(t_0 - \delta(t)) \cos k,$$

(4.74a)

$$(\mathbf{d}_\pm)_y = -\frac{1}{2}(t_0 - \delta(t)) \sin k,$$

(4.74b)

$$(\mathbf{d}_\pm)_z = \pm h_{st}(t).$$

(4.74c)

Thus, electrons with spin-up and spin-down are decoupled. It is noted that $(\mathbf{d}_\pm)_z$ differ by a minus sign. The corresponding Berry curvatures for electrons with spin-up and spin-down will also differ by a minus sign. As t increases from 0 to T, if an electron with spin-up moves from left to right, there must be another electron with spin-down moving from right to left:

$$\Delta P_\uparrow = +ea, \tag{4.75a}$$

$$\Delta P_\downarrow = -ea. \tag{4.75b}$$

As a result, there is no charge pump in a cyclic evolution. Instead electron spins at the ends exchange since electrons with spin-up and spin-down move in opposite directions simultaneously. When S_z is conserved, this idea can be used to describe the quantized spin pump.

Electron spin does not obey a fundamental conservation law. The concept of spin pump cannot be simply generalized to the case that S_z is non-conserved. However, Fu and Kane [11] proposed that similar physics happens even when the spin degrees of freedom are non-conserved. Consider the inclusion of an additional term for spin-orbit coupling,

$$V_{so} = \sum_{i,\sigma,\sigma'} i\,\mathbf{e}_{so} \cdot \sigma_{\sigma\sigma'} \left(c^\dagger_{i,\sigma}\sigma_{\sigma\sigma'}c_{i+1,\sigma'} - c^\dagger_{i+1,\sigma}\sigma_{\sigma\sigma'}c_{i,\sigma'} \right), \tag{4.76}$$

into Eq. (4.71), where $\mathbf{e}_{so}$ is an arbitrary vector characterizing the spin-orbit interaction. In this way, the z-component spin σ^z is no longer a good quantum number:

$$V_{so} = \sum_{i,\sigma,\sigma'} i\,\mathbf{e}_{so} \cdot \sigma_{\sigma\sigma'} \left[(1 - e^{ik})a^\dagger_{k,\sigma}b_{k,\sigma'} - (1 - e^{-ik})b^\dagger_{k,\sigma}a_{k,\sigma'} \right]. \tag{4.77}$$

In this case, there still exists an additional symmetry, that is, time reversal symmetry: the Hamiltonian satisfies the following relation:

$$H(-t) = \Theta H(t)\Theta^{-1}. \tag{4.78}$$

For an adiabatic cyclic evolution, we have

$$H(t) = H(t + T). \tag{4.79}$$

There exist two distinct points, $t_1^* = 0$ and $t_2^* = T/2$, at which the Hamiltonian is time reversal invariant:

$$H(t_i^*) = \Theta H(t_i^*)\Theta^{-1} \tag{4.80}$$

($i = 1, 2$). The existence of these two points plays a crucial role in the topological classification of the pump cycle.

In general, in the absence of a conservation law, there will be no level crossing, and the system will stay in the same state before and after cycling. In the case of charge pump, the level crossing is protected by the charge conservation. Here it is time reversal symmetry which protects the level crossing at t_1^* or t_2^*. At the two points, there exists a Kramers degeneracy: the two states as time reversal counterparts have the same energy. Fu and Kane proposed to introduce the concept of time reversal polarization, which is quantized in the spin pump.

4.7   Integer Quantum Hall Effect: Laughlin Argument

R.B. Laughlin showed the quantization of the Hall conductance as a consequence of the gauge invariance and the existence of the mobility gap [12]. Consider a two-dimensional electron gas, which is rolled as a cylinder along the y-direction as shown in Fig. 4.2. A magnetic flux ϕ is threading through the cylinder and varies with time very slowly. Suppose the system has an energy gap and the Fermi energy locates in the gap. According to the Faraday's law, the varying magnetic field induces an electric field E_y around the magnetic flux ϕ. The Hall current density J_x is given by

$$J_x = \sigma_{xy} E_y, \tag{4.81}$$

where the coefficient σ_{xy} is the Hall conductance. Then from the continuity condition of charge, the charge Q flowing through the cylinder is

$$\frac{dQ}{dt} = -\oint dl \cdot J_x = -\sigma_{xy} \oint dl \cdot E_y. \tag{4.82}$$

Using the Stokes' theorem,

$$\oint dl \cdot E_y = \iint dS \cdot \nabla \times E_y. \tag{4.83}$$

Furthermore, it follows from the Faraday's law, $\nabla \times \mathbf{E} = -\frac{\partial \mathbf{B}}{\partial t}$, that

$$\frac{dQ}{dt} = \sigma_{xy} \iint dS \cdot \frac{\partial \mathbf{B}}{\partial t} = \sigma_{xy} \frac{d\phi}{dt} \tag{4.84}$$

or

$$\Delta Q = \sigma_{xy} \Delta \phi, \tag{4.85}$$

where $\phi = \iint dS \cdot \mathbf{B}$ is the magnetic flux. Taking the change of magnetic flux as $\Delta \phi = \phi_0 = h/e$, the Hall conductance becomes $\sigma_{xy} = \frac{e}{h} \Delta Q$. Thus, the Hall

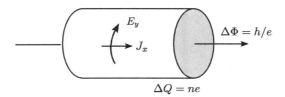

Fig. 4.2 A schematic of the setup for Laughlin's gedanken experiment for the integer quantum Hall effect. A changing flux through the cylindrical device generates an electric field E_y in the y-direction, which induces a Hall current J_x in the surface along the x-direction. The change of one quantum flux will transfer an integer of elementary charges from one side to the other side

conductance is determined by the charge transfer ΔQ after changing the magnetic flux by one magnetic flux quantum $\Delta \phi = \phi_0$.

What is the value of ΔQ? In the present geometry, the presence of the magnetic flux in the cylinder will lead to a gauge transformation in the vector potential:

$$\mathbf{p} + e\mathbf{A} \rightarrow \mathbf{p} + e(\mathbf{A} + \delta\mathbf{A}). \tag{4.86}$$

We take $\delta\mathbf{A} = \frac{\hbar}{e}\nabla\lambda$. The wave function will be transformed as

$$\Psi(r) \rightarrow e^{i\lambda(r)}\Psi(r). \tag{4.87}$$

For one quantum flux $\oint \delta\mathbf{A} \cdot dl = \phi_0$, one has $\lambda(\mathbf{r}, \phi = \phi_0) - \lambda(\mathbf{r}, \phi = 0) = 2\pi$. Thus, the eigenstates before and after the variation of one quantum flux are identical, that is,

$$H(\phi = \phi_0) = H(\phi = 0). \tag{4.88}$$

However, for a many-body system, the occupancy of electrons may be different after the variation of one quantum flux:

$$\Delta Q = ne, \tag{4.89}$$

where n is an integer and is determined by the topology of the band structure of the system. Therefore, we conclude that

$$\sigma_{xy} = n\frac{e^2}{h}. \tag{4.90}$$

This can be regarded as a generalization of the adiabatic charge pump in a two-dimensional system.

Fu-Kane argument is a spin version of Laughlin argument as a generalization from the integer quantum Hall effect to the quantum spin Hall effect, just like a generalization from charge pump to spin pump. For a quantum spin system, time reversal symmetry will give a different topological invariant for the quantum spin Hall system. Consider a setup of the same geometry as in the previous subsection for the quantum Hall effect as shown in Fig. 4.2. A magnetic flux ϕ threads a two-dimensional cylinder, which will cause an extra change of the phase factor before the physical states, $e^{i2\pi\phi/\phi_0}$. The magnetic flux plays the role of the edge crystal momentum k_x in the band theory. Increasing the magnetic flux with time t from $\phi = 0$ to ϕ_0 may form an adiabatic cyclic evolution. There exists a Kramers degeneracy at $\phi = 0$ and $\phi_0/2$:

$$H(0) = \Theta H(0)\Theta^{-1}, \tag{4.91a}$$

$$H(\phi_0/2) = \Theta H(\phi_0/2)\Theta^{-1}. \tag{4.91b}$$

Thus, variation by half flux quantum will change electron parity number at two ends.

4.8   Time Reversal Symmetry and the Z_2 Index

Time reversal symmetry implies that $[\mathcal{H}(\mathbf{r}), \Theta] = 0$, where the time reversal operator $\Theta = -i\sigma_y K$ and K is the complex conjugation. Note that in the band theory, time reversal symmetry means that

$$H(-\mathbf{k}) = \Theta H(\mathbf{k})\Theta^{-1}, \tag{4.92}$$

since the good quantum number $\mathbf{k}$ has already replaced the momentum operator $\mathbf{p} = -i\hbar\nabla$ in the Hamiltonian, and the later changes a minus sign under time reversal Θ. In the Brillouin zone of a square lattice, there are 4 (8 for a cubic lattice in three dimensions) time reversal invariant points satisfying $-\Gamma_i = \Gamma_i + n_i\mathbf{G}$, where $\mathbf{G}$ is a reciprocal lattice vector and $n_i = 0$ or 1 [11, 13, 17]. At these points, $\Gamma_i = n_i\mathbf{G}/2$:

$$H(\Gamma_i) = \Theta H(\Gamma_i)\Theta^{-1} \tag{4.93}$$

always holds; therefore, the eigenstates are always at least doubly degenerate due to the Kramers degeneracy. A pair of such energy bands $E_{2n-1}(k)$ and $E_{2n}(k)$ is called a Kramers pair, as illustrated in Fig. 4.3. These two bands (labeled as (n, I) and (n, II), respectively) are related to each other by time reversal operation accompanying with a phase factor [11]. Their crossings at time reversal invariant points are protected by time reversal symmetry. If a Kramers pair is isolated from other pairs by finite gaps, a topological invariant associated with this pair can be defined.

For simplicity, we consider a one-dimensional system and suppose that there is no additional degeneracy other than those required by time reversal symmetry. Therefore, the $2N$ eigenstates can be divided into N pairs that satisfy

$$\left| u_n^I(-k) \right\rangle = -e^{i\chi_{k,n}} \Theta \left| u_n^{II}(k) \right\rangle. \tag{4.94}$$

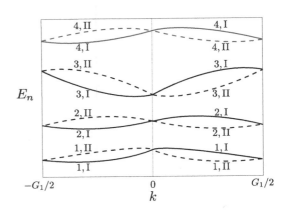

Fig. 4.3 Schematic of band structures $E_n(k)$ along the direction of one reciprocal vector. The Kramers pairs cross at time reversal invariant points $k = 0, G_1/2$

Then

$$\Theta \left| u_n^I(-k) \right\rangle = -\Theta e^{i\chi_{k,n}} \Theta \left| u_n^{II}(k) \right\rangle = e^{-i\chi_{k,n}} \left| u_n^{II}(k) \right\rangle \tag{4.95}$$

as $\Theta^2 = -1$ for electrons with spin $\frac{1}{2}$. Thus, one has the relation

$$\left| u_n^{II}(-k) \right\rangle = e^{i\chi_{-k,n}} \Theta \left| u_n^I(k) \right\rangle. \tag{4.96}$$

The partial polarization associated with one of the categories $s = I$ and II can be written as

$$P^s = \int_{BZ} \frac{dk}{2\pi} A_k^s, \tag{4.97}$$

with $A_k^s = i \sum_n \langle u_n^s(k) | \nabla_k | u_n^s(k) \rangle$. It is invariant (up to a lattice translation) under changes in the phases of $\left| u_n^I(k) \right\rangle$ and $\left| u_n^{II}(k) \right\rangle$. However, they appear to depend on the arbitrary choice of the label I and II assigned to each band. To make this invariance explicit for P^s, we separate the integral into two parts:

$$P^I = \int_0^\pi \frac{dk}{2\pi} A_k^I + \int_{-\pi}^0 \frac{dk}{2\pi} A_k^I,$$
$$= \int_0^\pi \frac{dk}{2\pi} A_k^I + \int_0^\pi \frac{dk}{2\pi} A_{-k}^I. \tag{4.98}$$

Using the time reversal constraint,

$$\left\langle \Theta u_n^{II}(k) | \partial_k | \Theta u_n^{II}(k) \right\rangle = - \left\langle u_n^{II}(k) | \partial_k | u_n^{II}(k) \right\rangle, \tag{4.99}$$

we have

$$A_{-k}^I = A_k^{II} - \sum_n \partial_k \chi_{k,n}. \tag{4.100}$$

It then follows that

$$P^I = \int_0^\pi \frac{dk}{2\pi} A_k - \frac{1}{2\pi} \sum_n (\chi_{\pi,n} - \chi_{0,n}), \tag{4.101}$$

where $A_k = A_k^I + A_k^{II}$. Introduce the $U(2N)$ matrix

$$w_{mn}(k) = \langle u_m(-k) | \Theta | u_n(k) \rangle. \tag{4.102}$$

Then only nonzero terms are

$$\left\langle u_n^I(-k) \right| \Theta \left| u_n^{II}(k) \right\rangle = -e^{-i\chi_{k,n}}, \tag{4.103a}$$
$$\left\langle u_n^{II}(-k) \right| \Theta \left| u_n^I(k) \right\rangle = e^{-i\chi_{-k,n}}. \tag{4.103b}$$

The matrix w is a direct product of 2×2 matrices with $-e^{-i\chi_{k,n}}$ and $e^{-i\chi_{-k,n}}$ on the off-diagonal. At $k = 0$ and π, w is antisymmetric. An antisymmetric matrix may be characterized by a Pfaffian, whose square is equal to the determinant. Then we have

$$\frac{\mathrm{Pf}\,[w(\pi)]}{\mathrm{Pf}\,[w(0)]} = \exp\left[i \sum_n (\chi_{\pi,n} - \chi_{0,n})\right]. \tag{4.104}$$

Thus,

$$P^I = \frac{1}{2\pi}\left[\int_0^\pi dk\, A_k + i\ln\frac{\mathrm{Pf}\,[w(\pi)]}{\mathrm{Pf}\,[w(0)]}\right]. \tag{4.105}$$

A similar formula can be obtained for P^{II}. It follows from the time reversal symmetry that $P^{II} = P^I$ modulo an integer, reflecting the Kramers pairing of the Wannier states. The charge polarization for one Kramers pair of states is

$$P_\rho = P^I + P^{II}, \tag{4.106}$$

and the time reversal polarization is defined as

$$\begin{aligned}
P_\theta &= P^I - P^{II} \\
&= \frac{1}{2\pi}\left[\int_0^\pi dk\, A_k - \int_{-\pi}^0 dk\, A_k + 2i\ln\frac{\mathrm{Pf}\,[w(\pi)]}{\mathrm{Pf}\,[w(0)]}\right].
\end{aligned} \tag{4.107}$$

In terms of the matrix w_{nm}, the formula can be written in a compact form:

$$P_\theta = \frac{1}{2\pi i}\left[\int_0^\pi dk\, \mathrm{Tr}[w^\dagger \nabla_k w] - 2\ln\frac{\mathrm{Pf}\,[w(\pi)]}{\mathrm{Pf}\,[w(0)]}\right]. \tag{4.108}$$

In the matrix w, only nonzero elements are off-diagonal:

$$\begin{aligned}
\mathrm{Tr}[w^\dagger \nabla_k w] &= \mathrm{Tr}\left[\begin{pmatrix} 0 & e^{-i\chi_{-k,n}} \\ -e^{i\chi_{k,n}} & 0 \end{pmatrix} \nabla_k \begin{pmatrix} 0 & -e^{-i\chi_{k,n}} \\ e^{i\chi_{-k,n}} & 0 \end{pmatrix}\right] \\
&= i\nabla_k \chi_{-k,n} - i\nabla_k \chi_{k,n}.
\end{aligned} \tag{4.109}$$

Using the unitarity of w, we have

$$\mathrm{Tr}[w^\dagger \nabla_k w] = \mathrm{Tr}[\nabla_k \ln w(k)] = \nabla_k \ln \det[w(k)]. \tag{4.110}$$

Thus, P_θ can be expressed as

$$P_\theta = \frac{1}{2\pi i}\left[\ln\frac{\det(w(\pi))}{\det(w(0))} - 2\ln\frac{\mathrm{Pf}(w(\pi))}{\mathrm{Pf}(w(0))}\right] \tag{4.111}$$

or

$$(-1)^{P_\theta} = \frac{\sqrt{\det(w(0))}}{\mathrm{Pf}(w(0))} \frac{\sqrt{\det(w(\pi))}}{\mathrm{Pf}(w(\pi))}. \tag{4.112}$$

In general, for a cyclic process of $H(t + T) = H(t)$, it follows that

$$H(t_1^* = 0) = \Theta H(0)\Theta^{-1}, \tag{4.113a}$$

$$H(t_1^* = T/2) = \Theta H(T/2)\Theta^{-1}. \tag{4.113b}$$

The change of time reversal polarization is gauge invariant

$$\nu = [P_\theta(T/2) - P_\theta(0)]\,\mathrm{mod}2. \tag{4.114}$$

Consider the mapping between the time reversal invariant momenta Γ_i and the time invariant point of time t_i^*; we conclude that the topological invariant can be written as

$$(-1)^\nu = \prod_i \frac{\sqrt{\det(w(\Gamma_i))}}{\mathrm{Pf}(w(\Gamma_i))}. \tag{4.115}$$

Since

$$\det(w(\Gamma_i)) = [\mathrm{Pf}(w(\Gamma_i))]^2, \tag{4.116}$$

the right-hand side of Eq. (4.115) is always $+1$ or -1. Correspondingly ν is only an integer modulo 2, that is, 0 or 1. Thus, the time reversal polarization defines two distinct polarization states, topologically trivial ($\nu = 0$) and nontrivial ($\nu = 1$). Fu and Kane proposed that the value of ν is related to the presence or the absence of a Kramers degenerate states at the end of a finite system [11].

If an insulator has the additional inversion symmetry, there is a simplified algorithm to calculate the Z_2 invariant. Suppose that the Hamiltonian H has an inversion symmetry,

$$H(-\mathbf{k}) = PH(\mathbf{k})P^{-1}, \tag{4.117}$$

where the parity operator is defined by

$$P\,|\mathbf{r}, s_z\rangle = P\,|-\mathbf{r}, s_z\rangle. \tag{4.118}$$

Here $\mathbf{r}$ is the coordinate and s_z is the spin which is unchanged by the parity P because it is a pseudovector. An explicit consequence of the combination of time reversal symmetry and inversion symmetry is the fact that the Berry curvature must vanish:

$$F(\mathbf{k}) = \nabla_\mathbf{k} \times \mathbf{A}(\mathbf{k}) = 0. \tag{4.119}$$

It follows from the definition of the Berry curvature that it is odd under time reversal, $F(-\mathbf{k}) = -F(\mathbf{k})$, and even under inversion, $F(-\mathbf{k}) = F(\mathbf{k})$. Considering the mth pair of the occupied energy bands at time reversal invariant momentum Γ_i, we define

$P |u_{2m,\Gamma_i}\rangle = \xi_{2m}(\Gamma_i) |u_{2m,\Gamma_i}\rangle$ where the parity eigenvalues $\xi_{2m}(\Gamma_i) = +1$ or -1. The degenerate Kramers partners share the same eigenvalue $\xi_{2m} = \xi_{2m-1}$. In this case, one has a simple formula to calculate δ [13]:

$$(-1)^\nu = \prod_i \prod_{m=1}^N \xi_{2m}(\Gamma_i). \tag{4.120}$$

In Sect. 3.6, we have already used this result to classify the topological phases in the lattice model.

Note on Pfaffian: In mathematics, a skew-symmetric matrix is a square matrix A whose transpose is its negative, $A = -A^T$. The determinant of a skew-symmetric matrix A can always be written as the square of a polynomial in the matrix entries, which is called the Pfaffian of the matrix, denoted by $\mathrm{Pf}(A)$, that is,

$$\det(A) = \mathrm{Pf}(A)^2. \tag{4.121}$$

The term Pfaffian was introduced by [15] who named it after Johann Friedrich Pfaff. The Pfaffian is nonvanishing only for $2n \times 2n$ skew-symmetric matrices, in which case it is a polynomial of degree n.

For example,

$$\mathrm{Pf}\begin{pmatrix} 0 & a \\ -a & 0 \end{pmatrix} = a, \tag{4.122}$$

$$\mathrm{Pf}\begin{pmatrix} 0 & a & b & c \\ -a & 0 & d & e \\ -b & -d & 0 & f \\ -c & -e & -f & 0 \end{pmatrix} = af - be + dc. \tag{4.123}$$

4.9 Generalization to Two and Three Dimensions

Generalization of the Z_2 invariant from two to three dimensions is a milestone in the development of topological insulator. The topological invariant characterizing a two-dimensional band structure may be constructed by rolling a two-dimensional system into a cylinder as shown in Fig. 4.4a. Then the magnetic flux threading the cylinder plays the role of the circumferential crystal momentum k_x, with $\phi = 0$ and $\phi = \phi_0/2$ corresponding to two edge time reversal momenta $k_x = \Lambda_1 = 0$ and $k_x = \Lambda_2 = \pi/a$. The Z_2 invariant characterizes the change of the time reversal polarization in the Kramers degeneracy at the ends of the one-dimensional system between $k_x = \Lambda_1$ and $k_x = \Lambda_2$. The change is related to the bulk band structure for a two-dimensional system with the periodic boundary condition. For a square lattice, there are four time reversal invariant momenta in the first Brillouin zone:

$$\Gamma_{n_x,n_y} = \left(\frac{n_x}{2}\mathbf{G}_x, \frac{n_y}{2}\mathbf{G}_y\right) \tag{4.124}$$

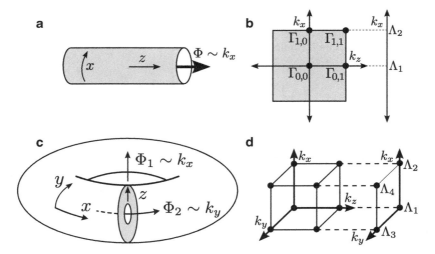

Fig. 4.4 (**a**) A two-dimensional cylinder threaded by magnetic flux Φ. When the cylinder has a circumference of a single lattice constant, Φ plays the role of the edge crystal momentum k_x in band theory. (**b**) The time reversal invariant fluxes $\Phi = 0$ and $h/2e$ correspond to edge time reversal invariant momenta $\Lambda_1 = 0$ and $\Lambda_2 = \pi/a$. Λ_a are projections of pairs of the four bulk time reversal momenta $\Gamma_{i=(a\mu)}$, which reside in the two-dimensional Brillouin zone indicated by the shaded region. (**c**) In three dimensions, the generalized cylinder can be visualized as a Corbino donut, with two fluxes, which correspond to the two components of the surface crystal momentum. (**d**) The four time reversal invariant fluxes $\Phi_1, \Phi_2 = 0, h/2e$ correspond to the four two-dimensional surface momenta (Reprinted with permission from [13]. Copyright (2007) by the APS)

with $n_x, n_y = 0, 1$. For an edge perpendicular to $\mathbf{G}_y$, the one-dimensional edge time reversal invariant momenta are $k_x = \Lambda_1$ and $k_x = \Lambda_2$, which satisfy $\Gamma_{1,n_y} - \Gamma_{0,n_y} = \frac{\mathbf{G}_x}{2}$. Thus, the time reversal polarization can be expressed as $\pi_x = \delta_{x1}\delta_{x2}$, where

$$\delta_{xi} = \frac{\sqrt{\det[w(\Gamma_{i,y})]}}{\text{Pf}[w(\Gamma_{i,y})]} = \pm 1. \tag{4.125}$$

However, π_x is not a gauge invariant. A k-dependent gauge transformation can change the sign of any pair of δ_i. If we roll the system into a cylinder along another direction, we can calculate the time reversal polarization $\pi_y = \delta_{y1}\delta_{y2}$. The product $\pi_x\pi_y$ is gauge invariant:

$$(-1)^\nu = \prod_{n_x,n_y=0,1} \frac{\sqrt{\det[w(\Gamma_{n_x,n_y})]}}{\text{Pf}[w(\Gamma_{n_x,n_y})]}. \tag{4.126}$$

This ν can be equal to 0 or 1 and define a single Z_2 invariant in two dimensions.

The Z_2 invariant for three-dimensional crystals can be reduced to the problems in two dimensions [11, 14, 16]. The three-dimensional Brillouin zone can be rolled into a donut along the x- and y-direction as illustrated in Fig. 4.4c. There are eight time reversal invariant momenta for three-dimensional systems:

$$\Gamma_{i=(n_1,n_2,n_3)} = \left(\frac{n_1}{2}\mathbf{G}_1, \frac{n_2}{2}\mathbf{G}_2, \frac{n_3}{2}\mathbf{G}_3 \right) \tag{4.127}$$

with $n_j = 0, 1$. They can be viewed as vertexes of a parallelepiped. For a fixed n_1, for example, $n_1 = 1$, the point set

$$\left(\frac{n_1}{2}\mathbf{G}_1, \frac{a_2}{2}\mathbf{G}_2, \frac{a_3}{2}\mathbf{G}_3 \right) \tag{4.128}$$

for all $a_2, a_3 \in [-\frac{1}{2}, \frac{1}{2})$ defines a two-dimensional Brillouin zone of a two-dimensional system respecting time reversal symmetry, for which a Z_2 invariant can be calculated from the method for two-dimensional system, referred as $\nu_{n_1=1}$. The other five invariants $\nu_{n_1=0}$, $\nu_{n_2=0,1}$, and $\nu_{n_3=0,1}$ can be defined in a similar way. These six invariants are associated with six planes of the above parallelepiped. Since they belong to the same three-dimensional crystal, only four of them are independent due to the constraints [11, 14]

$$\nu_{n_1=0} \cdot \nu_{n_1=1} = \nu_{n_2=0} \cdot \nu_{n_2=1} = \nu_{n_3=0} \cdot \nu_{n_3=1} \bmod 2. \tag{4.129}$$

The four independent invariants can be chosen as, say, $\nu_0 = \nu_{n_1=0}\nu_{n_1=1}$, $\nu_1 = \nu_{n_1=1}$, $\nu_2 = \nu_{n_2=1}$, and $\nu_3 = \nu_{n_3=1}$. The indices $\nu_0; (\nu_1\nu_2\nu_3)$ reflect the topology of the surface states [13, 17]. ν_0 is given by

$$(-1)^{\nu_0} = \prod_{n_1,n_2,n_3=0,1} \frac{\sqrt{\det[w(\Gamma_{n_1,n_2,n_3})]}}{\mathrm{Pf}[w(\Gamma_{n_1,n_2,n_3})]}. \tag{4.130}$$

If $\nu_0 = 1$, then the system is a strong topological insulator, with an odd number of Dirac cones on all surfaces of the crystal. If $\nu_0 = 0$, then the crystal is a weak topological insulator, with an even number (including 0) of Dirac cones at the surfaces. The latter one is topologically equivalent to a two-dimensional insulator and therefore is not robust against disorder. Let us take $0; (001)$ as an example [13]. The surface states corresponding to the two-dimensional Brillouin zone spanned by G_2 and G_3 (with index $\nu_1 = 0$) have two Dirac cones, and the same for the surface states in the Brillouin zone spanned by G_1 and G_3, with index $\nu_2 = 0$. But there is no surface states in G_2-G_3 plane, with index $\nu_3 = 1$.

4.10 Phase Diagram of Modified Dirac Equation

We come to study whether the modified Dirac equation is topologically trivial or nontrivial or not. The general solution of the wave functions for an infinite system or with the periodic boundary conditions can be expressed as

$$\Psi_\nu = u_\nu(\mathbf{p}) \exp[i\,(\mathbf{p} \cdot \mathbf{r} - E_{p,\nu} t)/\hbar], \tag{4.131}$$

in which the momentum is a good quantum number. The dispersion relations of four energy bands are

$$E_{p,\nu(=1,2)} = -E_{p,\nu(=3,4)} = \sqrt{v^2 p^2 + (mv^2 - Bp^2)^2}. \tag{4.132}$$

The four-component spinors $u_\nu(\mathbf{p})$ can be expressed as $u_\nu(\mathbf{p}) = S u_\nu(\mathbf{p} = 0)$ with

$$S = \sqrt{\frac{\epsilon_p}{2E_{p,1}}} \begin{pmatrix} 1 & 0 & -\frac{p_z v}{\epsilon_p} & -\frac{p_- v}{\epsilon_p} \\ 0 & 1 & -\frac{p_+ v}{\epsilon_p} & \frac{p_z v}{\epsilon_p} \\ \frac{p_z v}{\epsilon_p} & \frac{p_- v}{\epsilon_p} & 1 & 0 \\ \frac{p_+ v}{\epsilon_p} & -\frac{p_z v}{\epsilon_p} & 0 & 1 \end{pmatrix}, \tag{4.133}$$

where $p_\pm = p_x \pm i p_y$, $\epsilon_p = E_{p,1} + (mv^2 - Bp^2)$, and $u_\nu(0)$ is one of the four eigenstates of β.

The topological properties of the modified Dirac equation can be gained from these solutions. The Dirac equation is invariant under time reversal symmetry and can be classified according to the Z_2 topological classification following Kane and Mele [18]. In the representation for the Dirac matrices in Eq. (2.7a), the time reversal operator here is defined as $\Theta \equiv -i\alpha_x \alpha_z \mathcal{K}$ [19], where $\mathcal{K}$ is the complex conjugate operator that forms the complex conjugation of any coefficient that multiplies a ket or wave function (and stands on the right of $\mathcal{K}$). Under the time reversal operation, the modified Dirac equation remains invariant:

$$\Theta H(\mathbf{p}) \Theta^{-1} = H(-\mathbf{p}) \tag{4.134}$$

($\mathbf{p}$ is a good quantum number of the momentum). Furthermore, we have the relations that $\Theta u_1(\mathbf{p}) = -i u_2(-\mathbf{p})$ and $\Theta u_2(\mathbf{p}) = +i u_1(-\mathbf{p})$, which satisfy the relation of $\Theta^2 = -1$. Similarly, $\Theta u_3(\mathbf{p}) = -i u_4(-\mathbf{p})$ and $\Theta u_4(\mathbf{p}) = +i u_3(-\mathbf{p})$. Thus, the solutions of $\{u_1(\mathbf{p}), u_2(-\mathbf{p})\}$ and $\{u_3(\mathbf{p}), u_4(-\mathbf{p})\}$ are two degenerate Kramers pairs of positive and negative energies, respectively. The matrix of overlap $\{\langle u_\mu(\mathbf{p})| \, \Theta \, |u_\nu(\mathbf{p})\rangle\}$ has the form

$$
\begin{pmatrix}
0 & i\frac{mv^2-Bp^2}{E_{p,1}} & -i\frac{p-v}{E_{p,1}} & i\frac{p_z v}{E_{p,1}} \\
-i\frac{mv^2-Bp^2}{E_{p,1}} & 0 & i\frac{p_z v}{E_{p,1}} & i\frac{p+v}{E_{p,1}} \\
i\frac{p-v}{E_{p,1}} & -i\frac{p_z v}{E_{p,1}} & 0 & i\frac{mv^2-Bp^2}{E_{p,1}} \\
-i\frac{p_z v}{E_{p,1}} & -i\frac{p+v}{E_{p,1}} & -i\frac{mv^2-Bp^2}{E_{p,1}} & 0
\end{pmatrix}
\tag{4.135}
$$

which is antisymmetric, $\langle u_\mu(\mathbf{p}) | \Theta | u_\nu(\mathbf{p}) \rangle = -\langle u_\nu(\mathbf{p}) | \Theta | u_\mu(\mathbf{p}) \rangle$. For the two negative energy bands $u_3(\mathbf{p})$ and $u_4(\mathbf{p})$ which are fully occupied for an insulator, the submatrix of overlap can be expressed in terms of a single number as $\epsilon_{\mu\nu} P(\mathbf{p})$:

$$
P(\mathbf{p}) = i \frac{mv^2 - Bp^2}{\sqrt{(mv^2 - Bp^2)^2 + v^2 p^2}},
\tag{4.136}
$$

which is the Pfaffian for the 2×2 matrix. According to Kane and Mele [18], the even or odd number of the zeros in $P(\mathbf{p})$ defines the Z_2 topological invariant. Here we want to emphasize that the sign of a dimensionless parameter mB will determine the Z_2 invariant of the modified Dirac equation. Since $P(\mathbf{p})$ is always non-zero for $mB \leq 0$ and there exists no zero in the Pfaffian, we conclude immediately that the modified Dirac Hamiltonian for $mB \leq 0$ including the conventional Dirac Hamiltonian ($B = 0$) is topologically trivial.

For $mB > 0$, the case is different. In this continuous model, the Brillouin zone becomes infinite. At $p = 0$ and $p = +\infty$, $P(0) = i\,\mathrm{sgn}(m)$ and $P(+\infty) = -i\,\mathrm{sgn}(B)$. In this case, $P(\mathbf{p}) = 0$ at $p^2 = mv^2/B$. $\mathbf{p} = 0$ is always one of the time reversal invariant momenta. As a result of an isotropic model in the momentum space, we think all points of $p = +\infty$ shrink into one point if we regard the continuous model as a limit of the lattice model by taking the lattice space $a \to 0$ and the reciprocal lattice vector $G = 2\pi/a \to +\infty$. In this sense, as a limit of a square lattice, other three time reversal invariant momenta have $P(0, G/2) = P(G/2, 0) = P(G/2, G/2) = P(+\infty)$ which has an opposite sign of $P(0)$ if $mB > 0$. Similarly for a cubic lattice, $P(\mathbf{p})$ of other seven time reversal invariant momenta have opposite sign of $P(0)$. Following Fu, Kane, and Mele [13, 17], we conclude that *the modified Dirac Hamiltonian is topologically nontrivial only if $mB > 0$.*

In two dimensions, Z_2 index can be determined by evaluating the winding number of the phase of $P(p)$ around a loop, enclosing the half of the Brillouin zone in the complex plane of $\mathbf{p} = p_x + ip_y$:

$$
I = \frac{1}{2\pi i} \oint_C d\mathbf{p} \cdot \nabla_{\mathbf{p}} \log[P(\mathbf{p}) + i\delta].
\tag{4.137}
$$

Because the model is isotropic, the integral is then reduced to only the path along p_x-axis, while the part of the half-circle integral vanishes for $\delta > 0$ and $|\mathbf{p}| \to +\infty$.

Fig. 4.5 Phase diagram of topological states of the modified Dirac equation as a function of two model parameters m and B

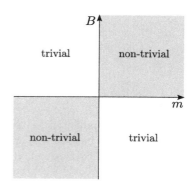

Along the p_x-axis one of a pair of zeros in the ring is enclosed in contour C when $mB > 0$, which gives a Z_2 index $\nu = 1$. This defines the nontrivial quantum spin Hall phase.

Volovik [20, 21] proposed that the Green's function rather than the Hamiltonian is more applicable to classify topological insulators. From the three-dimensional Dirac equation, the Green's function has the form

$$G(i\omega_n, \mathbf{p}) = \frac{1}{i\omega_n - H}$$

$$= -\frac{i\omega_n + v\mathbf{p} \cdot \alpha + (mv^2 - Bp^2)\beta}{\omega_n^2 + h^2(p)}, \qquad (4.138)$$

where $h^2(k) = H^2 = v^2 p^2 + (mv^2 - Bp^2)^2$. The frequency $\omega_n = (2n+1)\pi/\beta = (2n+1)\pi k_B T$ (k_B is the Boltzmann constant and T is the temperature). The topological invariant is defined as

$$\tilde{N} = \frac{1}{24\pi^2} \epsilon_{ijk} \text{Tr}[K \int\limits_{i\omega_n = 0} d\mathbf{p} G \partial_{p_i} G^{-1} G \partial_{p_j} G^{-1} G \partial_{p_k} G^{-1}], \qquad (4.139)$$

where $K = \sigma_y \otimes \sigma_0$ is the symmetry-related operator. After tedious algebra, it is found that

$$\tilde{N} = \text{sgn}(m) + \text{sgn}(B). \qquad (4.140)$$

When $mB > 0$, $\tilde{N} = \pm 2$, which defines the phase topologically nontrivial. If B is set to be positive, there exists a quantum phase transition from topologically trivial phase of $m < 0$ to a topologically nontrivial phase of $\mu > 0$. This is in a good agreement with the result of Z_2 index in the preceding section [22].

As a summary, we present the phase diagram according to the topological invariants which is presented in Fig. 4.5.

4.11 Further Reading

- L. Fu, C.L. Kane, Time reversal polarization and a Z_z adiabatic spin pump. Phys. Rev. B **74**, 195312 (2006)
- L. Fu, C.L. Kane, Topological insulators with inversion symmetry. Phys. Rev. B **76**, 045302 (2007)
- Q. Niu, D.J. Thouless, Y.-S. Wu, Quantized Hall conductance as a topological invariant. Phys. Rev. B **31**, 3372 (1985)
- D.J. Thouless, M. Kohmoto, M.P. Nightingale, M. den Nijs, Quantized Hall conductance in a two-dimensional periodic potential. Phys. Rev. Lett. **49**, 405 (1982)
- D. Xiao, M.C. Chang, Q. Niu, Berry phase effects on electronic properties. Rev. Mod. Phys. **82**, 1959 (2010)

References

1. C. Kittel, *Introduction to Solid State Physics*, 7th edn. (Willey, New York, 1996)
2. D. Xiao, M.C. Chang, Q. Niu, Rev. Mod. Phys. **82**, 1959 (2010)
3. M. Kohmoto, Ann. Phys. **160**, 343 (1985)
4. A. Messiah, *Quantum Mechanics*, Chap. XVII (Interscience, New York, 1961)
5. S.Q. Shen, Phys. Rev. B **70**, 081311(R) (2004)
6. M.C. Chang, Q. Niu, Phys. Rev. Lett. **75**, 1348 (1995)
7. G. Sundaram, Q. Niu, Phys. Rev. B **59**, 14915 (1999)
8. R. Resta, D. Vanderbilt, In: *Physics of Ferroelectrics: A modern Perspective*, ed. by K.M. Rabe, C.H. Ahn, J.M. Triscone (Springer, Berlin, 2007) p. 31.
9. R.D. King-Smith, D. Vanderbilt, Phys. Rev. B **47**, 1651 (1993)
10. M.J. Rice, E.J. Mele, Phys. Rev. Lett. **49**, 1455 (1982)
11. L. Fu, C.L. Kane, Phys. Rev. B **74**, 195312 (2006)
12. R.B. Laughlin, Phys. Rev. B **23**, 5632 (1981)
13. L. Fu, C.L. Kane, Phys. Rev. B **76**, 045302 (2007)
14. J.E. Moore, L. Balents, Phys. Rev. B **75**, 121306(R) (2007)
15. A. Cayley, Cambridge and Dublin Mathematical Journal VII, 40 (1852). Also see the item "Pfaffian" in Wikipedia
16. T. Fukui, Y. Hatsugai, J. Phys. Soc. Jpn. **76**, 053702 (2007)
17. L. Fu, C.L. Kane, E.J. Mele, Phys. Rev. Lett. **98**, 106803 (2007)
18. C.L. Kane, E.J. Mele, Phys. Rev. Lett. **95**, 146802 (2005)
19. J.D. Bjorken, S.D. Drell, *Relativistic Quantum Mechanics* (McGraw-Hill, New York, 1964), p. 72
20. G.E. Volovik, JETP Lett. **91**, 55 (2010)
21. G.E. *Volovik, The Universe in a Helium Droplet* (Clarendon, Oxford, 2003)
22. S.Q. Shen, W.Y. Shan, H.Z. Lu, SPIN **1**, 33 (2011)

Chapter 5
Topological Phases in One Dimension

Abstract Polyacetylene was extensively studied in the 1980s. Reexamination of Su-Schrieffer-Heeger model for polyacetylene shows that it is actually a one-dimensional topological insulator. Topological phases also exist in other one-dimensional systems.

Keywords One-dimensional topological insulator • Su-Schrieffer-Heeger model • p-wave pairing superconductor • Ising model • Maxwell's equation

5.1 Su-Schrieffer-Heeger Model for Polyacetylene

The simplest "two-band" model is the Su-Schrieffer-Heeger model for polyacetylene [1], which is an insulator with the chirality symmetry. Consider a one-dimensional dimerized lattice,

$$H = \sum_{n=1}^{N}(t + \delta t)c_{A,n}^{\dagger}c_{B,n} + \sum_{n=1}^{N-1}(t - \delta t)c_{A,n+1}^{\dagger}c_{B,n} + h.c., \tag{5.1}$$

where $c_{A(B),n}^{\dagger}$ and $c_{A(B),n}$ are the creation and annihilation operators of electron on A (or B) sublattice site $(A(B), n)$, respectively. In this model, each unit cell consists of two sites, A and B, and the hopping term connects the two different sublattice sites. The hopping amplitude in the unit cell is $t + \delta t$ and that between two unit cells is $t - \delta t$. There are two distinct phases named A and B phases which are plotted in Fig. 5.1. These two phases were believed to be degenerate. The interface of these two phases forms a domain wall, which may generate a soliton solution nearby. In this section, we demonstrate that these two phases are topologically distinct in the open boundary condition.

S.-Q. Shen, *Topological Insulators: Dirac Equation in Condensed Matters*,
Springer Series in Solid-State Sciences 174, DOI 10.1007/978-3-642-32858-9_5,
© Springer-Verlag Berlin Heidelberg 2012

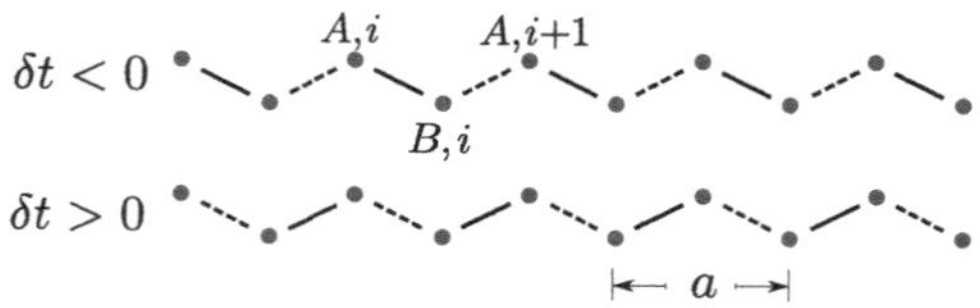

Fig. 5.1 Two distinct phases in Su-Schrieffer-Heeger model. The *solid* and *dashed lines* stand for the long and short bonds of hopping, respectively. Note that the boundary conditions are distinct in two phases

Performing the Fourier transformation,

$$a_k = \frac{1}{\sqrt{N}} \sum_n e^{-ik \cdot na} c_{A,n}, \qquad (5.2a)$$

$$b_k = \frac{1}{\sqrt{N}} \sum_n e^{-ik \cdot na} c_{B,n}, \qquad (5.2b)$$

where N is the number of the unit cells (the total number of lattice sites is $2N$), we obtain

$$H = (t + \delta t) \sum_{k \in (-\pi, \pi)} \left(a_k^\dagger b_k + b_k^\dagger a_k \right) + (t - \delta t) \sum_k \left(e^{ik} a_k^\dagger b_k + e^{-ik} b_k^\dagger a_k \right). \quad (5.3)$$

Introducing the spinor

$$\psi_k = \begin{pmatrix} a_k \\ b_k \end{pmatrix}, \qquad (5.4)$$

we can write the Hamiltonian in a compact form

$$H = \sum_k \psi_k^\dagger [((t + \delta t) + (t - \delta t) \cos k) \sigma_x + (t - \delta t) \sin k \sigma_y] \psi_k. \qquad (5.5)$$

Under a transformation, $\sigma_x \rightarrow \sigma_z, \sigma_y \rightarrow \sigma_x$, and $\sigma_z \rightarrow \sigma_y$ and $k \rightarrow k + \pi$, it is reduced to

$$H = \sum_k \psi_k^\dagger [-(t - \delta t) \sin k \sigma_x + (2\delta t + 2(t - \delta t) \sin^2 \frac{k}{2}) \sigma_z] \psi_k. \qquad (5.6)$$

Thus, a one-dimensional dimerized lattice is equivalent to the Dirac lattice model as we studied in Chap. 3.

In general, the dispersions of this two-band model are

$$E_\pm = \pm \sqrt{d_x^2 + d_z^2}, \qquad (5.7)$$

where $d_x = -(t - \delta t) \sin k$ and $d_z = 2\delta t + 2(t - \delta t) \sin^2 \frac{k}{2}$. The eigenstates for the negative dispersion are

$$|\varphi\rangle = \frac{1}{\sqrt{2}} \begin{pmatrix} \mathrm{sgn}(d_x) \sqrt{1 - \dfrac{d_z}{\sqrt{d_x^2 + d_z^2}}} \\ -\sqrt{1 + \dfrac{d_z}{\sqrt{d_x^2 + d_z^2}}} \end{pmatrix}. \tag{5.8}$$

They are fully filled for a half filling, that is, averagely one electron at every two sites. An energy gap $\Delta E = 4\delta t$ opens for $\delta t \neq 0$.

Thus, the Berry phase for this state is defined as

$$\begin{aligned}
\gamma &= \int_{-\pi}^{+\pi} dk \, \langle \varphi | \, i \, \partial_k \, | \varphi \rangle \\
&= \frac{1}{2} \int_{-\pi}^{+\pi} dk \, [i \, \partial_k \ln \mathrm{sgn}(d_x)] \left(1 - \frac{d_z}{\sqrt{d_x^2 + d_z^2}} \right) \\
&= \frac{1}{2} \int_{-\delta}^{+\delta} dk \, [i \, \partial_k \ln \mathrm{sgn}(d_x)] \, (1 - \mathrm{sgn}(\delta t)) \\
&\quad + \frac{1}{2} \int_{\pi-\delta}^{\pi+\delta} dk \, [i \, \partial_k \ln \mathrm{sgn}(d_x)] \, (1 - \mathrm{sgn}(t + \delta t)) \\
&= \frac{1}{2} \pi \, [\mathrm{sgn}(t + \delta t) - \mathrm{sgn}(\delta t)]
\end{aligned} \tag{5.9}$$

with a modulus 2π. For $\delta t > 0$, $\gamma = 0$, but for $\delta t < 0$, $\gamma = \pi$. This is consistent with the conclusion from the Dirac model. Alternatively, the winding index is given by

$$(-1)^\nu = \mathrm{sgn}(\delta t)\mathrm{sgn}(t + \delta t) = \mathrm{sgn}(1 + t/\delta t). \tag{5.10}$$

The change of the Berry phase or the winding number accompanies closing and reopening of the energy gap between the two bands near $\delta t = 0$. It can be regarded that the energy gap changes from positive to negative as shown in Fig. 5.2. At $\delta t = 0$, the spectrum is gapless and the two bands cross at $k = 0$. Near the point, using $\sin x \approx x$ for a small x, one obtains

$$H = \sum_k \psi_k^\dagger [-(t - \delta t)k\sigma_x + \left(2\delta t + \frac{1}{2}(t - \delta t)k^2 \right) \sigma_z] \psi_k. \tag{5.11}$$

This is the continuous model of the Dirac equation. Thus, we can define the energy gap $\Delta E = 4\delta t$, not $4|\delta t|$. The sign change of δt indicates an topological quantum phase transition.

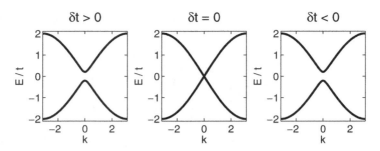

Fig. 5.2 The energy dispersions of $\delta t > 0$, $\delta t = 0$, and $\delta t < 0$. Closing and reopening of the energy gap near $\delta t = 0$ indicates occurrence of quantum phase transition

The existence of the end states in an open boundary condition is characteristic of the topological phase when the Berry phase is π or the winding index $\nu = 1$. It should be noted that the open boundary condition means that the chain is cut between two unit cells, not between two sites within a unit cell. Assume that $t > 0$. It is topologically nontrivial for $\delta t < 0$, but trivial for $\delta t > 0$. In other words, if the end bond is a long bond, $|t + \delta t| < |t - \delta t|$, it is topologically nontrivial. Otherwise it is topologically trivial.

A topological quantum transition occurs at $\delta t = 0$. In the long wave approximation, we can make use of the solution in Sect. 2.5.1 when $\delta t < 0$. In this case, there exists a solution of zero energy near the end. The spatial distribution of the wave function is mainly determined by the characteristic length

$$\xi_- = \frac{2\,|B|\,\hbar}{v}(1 - \sqrt{1 - 4mB})^{-1} \rightarrow \frac{\hbar}{|m|\,v} = \frac{t - \delta t}{2\,|\delta t|}. \tag{5.12}$$

It becomes divergent when $\delta t \rightarrow 0$, which illustrates that the end state evolves into a bulk one and the system becomes gapless. There is no end state when $\delta t > 0$. Therefore, the fact demonstrates a topological quantum phase near $\delta t = 0$ [2].

We can also use numerical method to calculate the energy eigenstates and eigenvalues by diagonalizing the Hamiltonian, which can be written in the form of square matrix

$$H = \begin{pmatrix} 0 & t + \delta t & 0 & 0 & 0 & 0 & 0 \\ t + \delta t & 0 & t - \delta t & 0 & 0 & 0 & 0 \\ 0 & t - \delta t & 0 & t + \delta t & 0 & 0 & 0 \\ 0 & 0 & t + \delta t & 0 & t - \delta t & 0 & 0 \\ 0 & 0 & 0 & t - \delta t & 0 & \ddots & 0 \\ 0 & 0 & 0 & 0 & \ddots & 0 & t + \delta t \\ 0 & 0 & 0 & 0 & 0 & t + \delta t & 0 \end{pmatrix}. \tag{5.13}$$

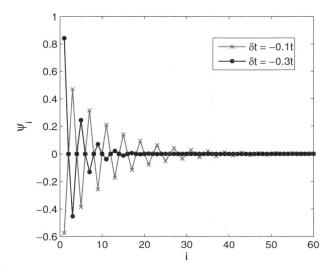

Fig. 5.3 The amplitudes of the wave function Ψ_i of the end states at the lattice site i for two different $\delta t = -0.1t$ and $-0.3t$. The smaller value of $|\delta t|$ corresponds to the wider distribution of the wave function in space

One can find the existence of the zero-energy mode at the end while changing the sign of δt. The end state solutions at different $\delta t = -0.1t$ and $-0.3t$ are plotted in Fig. 5.3. It demonstrates that the wave function has a wider distribution in the space for a small $|\delta t|$.

However, the most famous excitations in this model are soliton and antisoliton, which are charge and spin carriers in polyacetylene [3]. They are the domain walls of two distinct phases of π and 0. These solutions correspond to those of the Dirac equation at the interface between two regions of positive and negative masses in Chap. 2. The wave function of the in-gap bound state is distributed around the domain walls. Considering the degeneracy of electron spins, there are two bound states with different spins. The charge and spin states of the soliton are followed from the solutions of domain wall along with the localized chemical-bond representation. Totally there are four possible states according to the electron number n in the two states: (a) two neutral spin-$\frac{1}{2}$ solitons with $S_z = \pm\frac{1}{2}$ for $n = 1$ and (b) two charge species $S^\pm$ for $n = 0$ and $n = 2$, in which the total spin is zero and may be viewed as spinless "ions." However, the solitons can move freely unless they are pinned, in contrast to the chemical analogs. From the point of view of topological insulator, these states are the end states at the interface between one topologically trivial phase and one topologically nontrivial phase.

5.2 Ferromagnet with Spin-Orbit Coupling

In the Su-Schrieffer-Heeger model, the Hamiltonian is written as a 2×2 matrix in the basis of A and B sublattices, $(a_k^\dagger, b_k^\dagger)$. A new type of topological phase in one dimension can be obtained if we replace the basis by electrons with different spins, $(c_{k,\uparrow}^\dagger, c_{k,\downarrow}^\dagger)$. For a ferromagnet with spin-orbit coupling, one yields

$$H = \sum_k (c_{k,\uparrow}^\dagger, c_{k,\downarrow}^\dagger) \left[\lambda \sin k \sigma_x + \left(M - 4B \sin^2 \frac{k}{2} \right) \sigma_z \right] \begin{pmatrix} c_{k,\uparrow} \\ c_{k,\downarrow} \end{pmatrix}, \qquad (5.14)$$

where $c_{k,\sigma}^\dagger$ and $c_{k,\sigma}$ are the creation and annihilation operators for electrons with spin $\sigma(=\uparrow, \downarrow)$. Here λ is the strength of spin-orbit coupling. In the absence of the spin-orbit coupling, the two bands of electrons with spin-up and spin-down are well separated. If the lower band is fully filled, the ground state is fully saturated with a maximal spin and the system is an insulating ferromagnet. In the presence of the spin-orbit coupling λ, the S_z is no longer conserved. However, the filled band is still ferromagnetic as the expectation value of S_z is still nonzero. We find that this model has the identical mathematical structure as the Su-Schrieffer-Heeger model, although the bases for the two models are different.

5.3 *p*-Wave Pairing Superconductor

The *p*-wave pairing spinless superconductor has two distinct phases, strong pairing and weak pairing, which correspond to the topologically trivial and nontrivial phase, respectively [4]. In the Bardeen-Cooper-Schrieffer theory for superconductivity, the effective model for the superconductor can be written as

$$H = \sum_k \left(\frac{\hbar^2 k^2}{2m} - \mu \right) c_k^\dagger c_k + \Delta_k c_k c_{-k} + h.c., \qquad (5.15)$$

where μ is the chemical potential to determine the number of electrons. Introducing the spinor $(c_k^\dagger, c_{-k})$, one obtains

$$H = \sum_k (c_k^\dagger, c_{-k}) \left[\Delta_k \sigma_x + \frac{1}{2} \left(\frac{\hbar^2 k^2}{2m} - \mu \right) \sigma_z \right] \begin{pmatrix} c_k \\ c_{-k}^\dagger \end{pmatrix}. \qquad (5.16)$$

Here a constant $\frac{1}{2} \sum_k (\frac{\hbar^2 k^2}{2m} - \mu)$ is omitted. For a *p*-wave pairing superconductor, the order parameter for the Cooper pairing satisfies $\Delta_k = -\Delta_{-k}$. For a simplicity, here we take $\Delta_k = \Delta_0 k$. The Berry phase in the ground state is always π for $\mu > 0$ as m is assumed to be positive. In this system, if $\Delta = 0$,

$$H = \frac{1}{2} \sum_k (c_k^\dagger, c_{-k}) \left(\frac{\hbar^2 k^2}{2m} - \mu \right) \sigma_z \begin{pmatrix} c_k \\ c_{-k}^\dagger \end{pmatrix}$$

$$= \sum_k \left(\frac{\hbar^2 k^2}{2m} - \mu \right) c_k^\dagger c_k - \frac{1}{2} \sum_k \left(\frac{\hbar^2 k^2}{2m} - \mu \right), \qquad (5.17)$$

the two states with eigenvalues $\pm \frac{1}{2}(\frac{\hbar^2 k^2}{2m} - \mu)$ actually correspond to one state. This is because the basis in the term of spinors is redundant. This so-called particle-hole symmetry persists even when $\Delta \neq 0$.

On a lattice, k and k^2 can be replaced by $\sin k$ and $4 \sin^2 \frac{k}{2}$. The effective model is

$$H = \sum_k (c_k^\dagger, c_{-k}) \left[\Delta_0 \sin k \sigma_x + \left(t + 4t' \sin^2 \frac{k}{2} \right) \sigma_z \right] \begin{pmatrix} c_k \\ c_{-k}^\dagger \end{pmatrix}. \qquad (5.18)$$

The energy eigenvalues of the quasiparticles always appear in pairs,

$$E_{\pm,k} = \pm \sqrt{\Delta_0^2 \sin^2 k + \left(t + 4t' \sin^2 \frac{k}{2} \right)^2}. \qquad (5.19)$$

Performing the Fourier transformation, one obtains a lattice model in the real space, which is the one-dimensional Kitaev model for Majorana fermion.

When the system has an open boundary condition, for a topologically nontrivial phase, there exists an energy zero mode near the boundary, which satisfies

$$\gamma^\dagger(E = 0) = \gamma(E = 0). \qquad (5.20)$$

Thus, the creation operator of the zero mode is equal to its annihilation operator. This particle is called Majarona fermion. Because of the particle-hole symmetry, these two states are actually one state after the particle-hole transformation. Thus, the ground states are doubly degenerate depending on whether the zero-energy mode is occupied or not. Since the Cooper pairing term in the effective Hamiltonian creates or annihilates the electrons in pairs, the number parity of electrons is always conserved. The occupancy of the zero mode changes the number parity of the system.

The p-wave pairing superconductor and the Su-Schrieffer-Heeger model are connected through a partial particle-hole transformation [5]. Performing a particle-hole transformation for electrons on the site B,

$$c_{B,n} \rightarrow c_{B,n}^\dagger$$

the Su-Schrieffer-Heeger model in Eq. (5.1) is transformed into

$$H = \sum_{n=1}^{N} (t + \delta t) c_{A,n}^\dagger c_{B,n}^\dagger + \sum_{n=1}^{N-1} (t - \delta t) c_{A,n+1}^\dagger c_{B,n}^\dagger + h.c.,$$

which is one for the p-wave pairing superconductor on a lattice [6]. A solution for the end states of the lattice model in terms of Majorana fermions can be found in Sect. 10.2.2.

5.4 Ising Model in a Transverse Field

The one-dimensional Ising model is equivalent to a spinless p-wave pairing super-conductor under the Jordan-Wigner transformation. The transverse Ising model is defined as

$$H = J \sum_{n=1}^{N-1} \sigma_{x,n}\sigma_{x,n+1} + h \sum_{n=1}^{N-1} \sigma_{z,n} \tag{5.21}$$

where N is the number of lattice sites.

When $|J| >> |h|$, the ground state is determined by the first term. It is antiferromagnetic if $J > 0$, and ferromagnetic if $J < 0$. The magnetization is along the x-direction, and the ground state is doubly degenerate. If $|h| >> |J|$, the ground state is ferromagnetic along the z-direction and is non-degenerate. Thus, the change of the degeneracy of the ground state reveals a quantum phase transition at $J = h$.

For a lattice with even number lattice sites, under the Jordan-Wigner transformation [7],

$$\sigma_n^+ = \sigma_{x,n} + i\sigma_{y,n} = 2 \exp[-i\pi \sum_{k=1}^{n-1} f_k^\dagger f_k] f_n^\dagger, \tag{5.22a}$$

$$\sigma_n^- = \sigma_{x,n} - i\sigma_{y,n} = 2 \exp[+i\pi \sum_{k=1}^{n-1} f_k^\dagger f_k] f_n^\dagger, \tag{5.22b}$$

$$\sigma_{z,n} = 2 f_n^\dagger f_n - 1, \tag{5.22c}$$

where $f_n^\dagger$ and f_n are the fermion operators and satisfy the anticommutation relation of $\{f_n^\dagger, f_{n'}\} = \delta_{n,n'}$. In this way, the model is reduced into one for a p-wave pairing superconductor or the Kitaev's toy model for Majorana fermion,

$$H = J \sum_{n=1}^{N-1} (f_n^\dagger - f_n)\left(f_{n+1}^\dagger + f_{n+1}\right) + h \sum_{n=1}^{N} (2 f_n^\dagger f_n - 1). \tag{5.23}$$

The ground state is doubly degenerate due to the existence of the end states when $J < h$. However, it is noted that the Jordan-Wigner transformation is not a local transformation. The ground states in the Ising model simply have different polarizations along the x-direction, not the end states.

5.5 One-Dimensional Maxwell's Equations in Media

A one-dimensional plane electromagnetic wave of the frequency ω in a nonconducting media can be described by the Maxwell's equations [8],

$$\partial_x E_y = -i\omega\mu(x)H_z, \tag{5.24a}$$

$$\partial_x H_z = -i\omega\epsilon(x)E_y. \tag{5.24b}$$

E_y is the electric field, and $H_z = \frac{1}{\mu_0}B_z - M_z$ is the magnetic field. ϵ is the electric permittivity, and μ is the magnetic permeability, which are a function of position.

To derive a Dirac-like equation for the electromagnetic wave, we introduce dimensionless fields, $e = E_y/E_0$ and $h = H_z/H_0$ where E_0 and H_0 are the field as units for the electric and magnetic fields, respectively, and can be determined by the incident wave, that is, $E_0/H_0 = 1/c\epsilon_0 = c\mu_0$ in the vacuum. Equation (5.24a and b) can be combined to write in the form of matrix,

$$-i\sigma_x\partial_x\begin{pmatrix} e \\ h \end{pmatrix} = \left(-\frac{k}{2}(\tilde{\mu} + \tilde{\epsilon}) + \frac{k}{2}(\tilde{\mu} - \tilde{\epsilon})\sigma_z\right)\begin{pmatrix} e \\ h \end{pmatrix}, \tag{5.25}$$

where $k = \omega/c$, $\tilde{\epsilon}(x) = \epsilon(x)/\epsilon_0$, and $\tilde{\mu} = \mu(x)/\mu_0$. The dimensionless permittivity $\tilde{\epsilon}(x)$ and permeability $\tilde{\mu}(x)$ satisfy the relation $\tilde{\epsilon}\tilde{\mu} = n(x)$ where $n(x)$ is the index of refraction. In this way, we obtain a Dirac-like equation for the electromagnetic wave

$$[-i\sigma_x\partial_x + m(x)\sigma_z + V(x)]\begin{pmatrix} e \\ h \end{pmatrix} = E\begin{pmatrix} e \\ h \end{pmatrix}. \tag{5.26}$$

Here the mass distribution $m(x) = \frac{k}{2}(\tilde{\epsilon} - \tilde{\mu})$, and the potential $V(x) - E = \frac{k}{2}(\tilde{\epsilon} + \tilde{\mu})$. This equation looks like the stationary Dirac equation with the eigenvalue E ($\hbar = c = 1$).

In the metamaterial with subwavelength resonant unit cells, both ϵ and μ can be tuned and even change their signs [9]. From this equation, it is possible to simulate the topological phase by using the microwave experiment in metamaterials. For example, design a sample with an interface with $m(x) > 0$ if $x > 0$ and $m(x) < 0$ if $x < 0$. It is required that $E = V(x) = 0$. It follows from Eq. (5.26) that we may have a solution which is distributed around the interface as that for a domain wall as shown in Fig. 2.1. Furthermore, if we design a periodic structure for $m(x)$, it is possible to have the solution for the end states as we plotted in Fig. 2.3. In this way, the topological phase can be observed in quasi-one-dimensional periodic metamaterial. This provides a platform to observe the topological excitations in one dimension.

5.6 Summary

Reexamination of Su-Schrieffer-Heeger model tells that polyacetylene actually has two distinct topological phases. The domain wall of these two phases constitutes the topological excitations or charge and spin carriers in the system. Also the Dirac equation in different bases may be applied to describe topological phases in different physical systems such as a dimerized lattice model, ferromagnet with spin-orbit coupling, and superconductor.

References

1. W.P. Su, J.R. Schrieffer, A.J. Heeger, Phys. Rev. Lett. **42**, 1698 (1979)
2. H.M. Guo, S.Q. Shen, Phys. Rev. B **84**, 195107 (2011)
3. A.J. Heeger, S. Kivelson, J.R. Schrieffer, W.P. Su, Rev. Mod. Phys. **60**, 781 (1988)
4. N. Read, D. Green, Phys. Rev. B **61**, 10267 (2000)
5. F.C. Pu, S.Q Shen, Phys. Rev. B **50**, 16086 (1994)
6. A. Kitaev, Phys. Usp. **44**, 131 (2001)
7. P. Jordan, E. Wigner, Z. Phys. **47**, 631 (1928)
8. J.D. Jackson, *Classical Electrodynamics*, 3rd edn. (Wiley, New York, 1999).
9. H.T. Jiang, Y.H. Li, Z.G. Wang, Y.W. Zhang, H. Chen, Philos. Mag. **92**, 1317 (2012)

Chapter 6
Quantum Spin Hall Effect

Abstract A quantum spin Hall system possesses a pair of helical edge states. It exhibits the quantum spin Hall effect, in which an electric current can induce a transverse spin current or spin accumulation near the system boundary.

Keywords Two-dimensional topological insulator • Chern number • Quantum Hall conductance • Quantum spin Hall effect • Helical edge states • Landauer-Büttiker formalism • HgTe/CdTe quantum well

6.1 Two-Dimensional Dirac Model and the Chern Number

In two dimensions, the Chern number is associated with the quantum Hall conductance in the band insulators. Before we introduce the quantum spin Hall effect, we first focus on the Chern number in two-dimensional Dirac equation in Eq. (2.42), in which time reversal symmetry is broken. The Hamiltonian can be written in a compact form,

$$H = \mathbf{d}(\mathbf{p}) \cdot \sigma, \tag{6.1}$$

where $d_x = v p_x$, $d_y = v p_y$, and $d_z = m v^2 - B p^2$. Using the formula in Eq. (A.29), the Chern number is given by

$$n_c = -\frac{1}{2} \left(\mathrm{sgn}(m) + \mathrm{sgn}(B) \right). \tag{6.2}$$

From this formula, we have two topological nontrivial phases with $n = \pm 1$ for $mB > 0$ and topologically trivial phase with $n = 0$ for $mB < 0$. We also have two marginal phases with $n = \pm \frac{1}{2}$ for $m = 0$ or $B = 0$. The massive Dirac fermions of $B = 0$ are a marginal phase. At the junction of two systems of a positive mass and a negative mass, the topological invariant changes by $\delta n = 1$ or -1. Thus, there exists a boundary state at the junction. For the gapless Dirac fermions $m = 0$ and $B \neq 0$, the system is also marginal. The topological invariant also changes by $\delta n = 1$ at the interface between positive and negative B.

In the lattice model in Eq. (3.31), using the formula for the Chern number in Eq. (A.29), one obtains

$$n_c = \begin{cases} 1 & \text{if } 0 < \Delta/B < 4 \\ -1 & \text{if } 4 < \Delta/B < 8 \end{cases}. \tag{6.3}$$

The number is always an integer as the first Brillouin zone is finite for a lattice model. There exist three transition points: the first transition point at $\Delta/B = 0$ is from $n_c = 0$ to $n_c = 1$; the second point at $\Delta/B = 4$ is from $n_c = 1$ to $n_c = -1$; and the third point at $\Delta/B = 8$ is from $n_c = -1$ to $n_c = 0$. It is noted that the transition at $\Delta/B = 4$ is between two topological phases with $n_c = 1$ and -1.

Nonzero Chern number indicates the quantum Hall conductance. Therefore the two-dimensional Dirac equation is a good candidate to study the quantum anomalous Hall effect in ferromagnetic insulator with spin-orbit coupling.

6.2 From Haldane Model to Kane-Mele Model

6.2.1 Haldane Model

In 1988, Haldane proposed a spinless fermion model for the integer quantum Hall effect without Landau levels, in which two independent effective Hamiltonians in the same form of two-dimensional Dirac equation were obtained [1]. He proposed that the quantum Hall effect may result from the broken time reversal symmetry without any net magnetic flux through the unit cell of a periodic two-dimensional graphite or graphene model as depicted in Fig. 6.1. The lattice is bipartite with A (black) and B (white) sublattice sites. A real hopping term t_1 between the nearest neighbor sites (solid line) and t_2 between the next nearest neighbor sites (dashed line) are considered. The on-site energy $+M$ on A sites and $-M$ on B sites are included to break the inversion symmetry on A and B sublattices. Besides, he added

Fig. 6.1 The Haldane honeycomb model. The *white* and *black dots* represents the two sublattice sites with different on-site energy. The areas a and b are threaded by the magnetic flux ϕ_a and $\phi_b = -\phi_a$, respectively. The area c has no flux

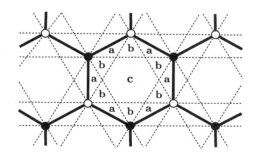

a periodic magnetic flux density $\mathbf{B}(r)$ normal to the plane with the full symmetry of the lattice and with the zero total flux through the unit cell, that is, the flux ϕ_a in the region a and the flux ϕ_b in the region b have the relation $\phi_a = -\phi_b$. Since the closed path of the nearest neighbor hops enclose complete unit cell, but the net flux is zero, the hopping terms t_1 are not affected, but the hopping terms t_2 acquire a phase $\phi = 2\pi(2\phi_a + \phi_b)/\phi_0$ where the flux quantum $\phi_0 = h/e$. The hopping direction is for which the amplitudes are $t_2 \exp[i\phi]$.

To diagonalize the Hamiltonian, a two-component spinor $\left(c_{k,A}^\dagger, c_{k,B}^\dagger\right)$ of Bloch states constructed on the two sublattices is applied. Let $(\mathbf{a}_1, \mathbf{a}_2, \mathbf{a}_3)$ be the displacements from a B site to its three adjacent A sites. In this representation, the model Hamiltonian can be expressed as

$$H = \epsilon(k) + \mathbf{d}(k) \cdot \sigma \tag{6.4}$$

where

$$\epsilon(k) = 2t_2 \cos\phi \sum_{i=1,2,3} \cos(\mathbf{k} \cdot \mathbf{b}_i), \tag{6.5a}$$

$$d_x(k) = +t_1 \sum_{i=1,2,3} \cos(\mathbf{k} \cdot \mathbf{a}_i), \tag{6.5b}$$

$$d_y(k) = +t_1 \sum_{i=1,2,3} \sin(\mathbf{k} \cdot \mathbf{a}_i), \tag{6.5c}$$

$$d_z(k) = M - 2t_2 \sin\phi \left(\sum_{i=1,2,3} \sin(\mathbf{k} \cdot \mathbf{b}_i) \right) \tag{6.5d}$$

with $\mathbf{b}_1 = \mathbf{a}_2 - \mathbf{a}_3$, $\mathbf{b}_2 = \mathbf{a}_3 - \mathbf{a}_1$, and $\mathbf{b}_3 = \mathbf{a}_1 - \mathbf{a}_2$. The Brillouin zone is a hexagon rotated by $\pi/2$ with respect to the Wigner-Seitz unit cell. At its six corners $(\mathbf{k} \cdot \mathbf{a}_1, \mathbf{k} \cdot \mathbf{a}_2, \mathbf{k} \cdot \mathbf{a}_3)$ is a permutation of $(0, 2\pi/3, 4\pi/3)$. Two distinct corners $\mathbf{k}_\alpha^0$ are defined such that $\mathbf{k}_\alpha^0 \cdot b_i = \alpha\frac{2\pi}{3}$ with $\alpha = \pm 1$. The energy spectra are easily obtained by diagonalizing the 2×2 matrix. There are two bands, which only touch if all three Pauli matrix terms have vanishing coefficients. This can only occur at the zone corner $\mathbf{k}_\alpha^0$ and only if $M = \alpha 3\sqrt{3}t_2 \sin\phi$. Assume $|t_2/t_1| < 1/3$, which guarantees that the two bands never overlap and are separated by a finite gap unless they touch. At the point $\mathbf{K}$, $(\mathbf{K} \cdot \mathbf{a}_1, \mathbf{K} \cdot \mathbf{a}_2, \mathbf{K} \cdot \mathbf{a}_3) = (0, 2\pi/3, -2\pi/3)$. Near the point, we expand the Hamiltonian to the linear order in $\delta\mathbf{k} = \mathbf{k} - \mathbf{K}$. As a result,

$$H_+ = v(\delta k_x \sigma_x - \delta k_y \sigma_y) + m_+ v^2 \sigma_z \tag{6.6}$$

where $v = \frac{3}{2}t_1 a/\hbar$ and $m_+ v^2 = M - 3\sqrt{3}t_2 \sin\phi$. At another point $\mathbf{K}'$, $(\mathbf{K}' \cdot \mathbf{a}_1, \mathbf{K}' \cdot \mathbf{a}_2, \mathbf{K}' \cdot \mathbf{a}_3) = (0, -2\pi/3, +2\pi/3)$,

$$H_- = v(-\delta k_x \sigma_x - \delta k_y \sigma_y) + m_- v^2 \sigma_z \tag{6.7}$$

where $v = \frac{3}{2}t_1 a/\hbar$ and $m_- v^2 = M + 3\sqrt{3}t_2 \sin\phi$. The two Hamiltonians have different chirality when $m_\pm = 0$.

To compare H_+ and H_-, we make a transformation for σ in H_-,

$$(\sigma_x, \sigma_y, \sigma_z) \rightarrow (-\sigma_x, \sigma_y, -\sigma_z). \tag{6.8}$$

Thus, H_- can be written as

$$H_- = v(\delta k_x \sigma_x - \delta k_y \sigma_y) + \tilde{m}_- v^2 \sigma_z \tag{6.9}$$

with $\tilde{m}_- = -m_- = -M - 3\sqrt{3}t_2 \sin\phi$. Therefore, the effective models near the two points have the form

$$H_\alpha = v(\delta k_x \sigma_x - \delta k_y \sigma_y) + \tilde{m}_\alpha v^2 \sigma_z \tag{6.10}$$

where $\tilde{m}_\alpha = \alpha M - 3\sqrt{3}t_2 \sin\phi$. Clearly, inclusion of M in the graphene lattice opens opposite energy gaps M and $-M$ at $\mathbf{K}$ and $\mathbf{K}'$, respectively while the magnetic flux opens the same energy gap at the two points. This demonstrates that the on-site energy $\pm M$ and magnetic flux play different roles in opening of energy gap and generate different topological results.

The Chern number of the whole system is determined by

$$n_c = \frac{1}{2} \left[\text{sgn}(\tilde{m}_+) + \text{sgn}(\tilde{m}_-) \right]. \tag{6.11}$$

In the absence of magnetic flux, the Chern number is always zero as the gaps at $\mathbf{K}$ and $\mathbf{K}'$ have opposite signs, while it can be $+1$ or -1 possibly in the presence of magnetic flux.

Of course, the topology of the system should be determined by the whole band structure. In his pioneering paper, Haldane used the Streda formula to calculate the Hall conductance [2],

$$\sigma_H = \left. \frac{\partial \rho}{\partial B_z} \right|_\mu, \tag{6.12}$$

the variation of density of charge carriers ρ with respect to the external field B_z perpendicular to the plane for a fixed chemical potential μ. For a full and complete calculation, we can use the formula in Eq. (A.29) to calculate the Chern number, which gives $+1$, 0 or -1. The Hall conductance is expressed in terms of the Chern number $\sigma_{xy} = n_c \frac{e^2}{h}$, which depicts a phase diagram as in Fig. 6.2.

6.2.2 Kane-Mele Model

The Haldane model is for spinless fermions. One can generalize the Haldane model to an electron system with spin, which becomes doubly degenerate if there is no

Fig. 6.2 The phase diagram of the Haldane model for $|t_2/t_1| < 1/3$. The zero-field quantum Hall effect phases where $\sigma_{xy} = n_c e^2 / h$ occur if $|M/t_2| < 3\sqrt{3}\,|\sin\phi|$

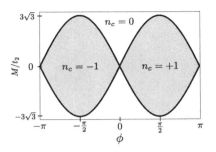

coupling between electrons with spin-up and spin-down. The electrons in the edge channel are chiral, that is, flowing around the boundary in the counterclockwise for $n_c = 1$ and in the clockwise for $n_c = -1$, which is characteristic of quantum Hall effect. This is a trivial generalization. In 2005, Kane and Mele [3] generalized the Haldane model to the graphene lattice model of electrons with spin $\frac{1}{2}$. They introduced the spin-orbit coupling between electron spin and momentum to replace the periodic magnetic flux and predicted a new quantum phenomenon – the quantum spin Hall effect. Simply speaking, the quantum spin Hall effect can be regarded as the combination of two layers of the Haldane models for electrons with spin-up and spin-down. In a system with time reversal symmetry, the electrons with spin-up in the edge channel flow in one direction, while electrons with spin-down in the edge channel flow in an opposite direction, $I_\uparrow = -I_\downarrow$. The net charge current in two edge channels is zero, $I_c \equiv I_\uparrow + I_\downarrow = 0$ as a net charge current breaks time reversal symmetry. Instead, a pure spin current circulates around the boundary of system, $I_s \equiv \frac{\hbar}{2e}(I_\uparrow - I_\downarrow)$. Unlike the quantum Hall effect in which the magnetic field breaks time reversal symmetry, the spin-orbit coupling preserves time reversal symmetry. The spin current itself does not break the symmetry since the momentum $p \to -p$ and spin $\sigma \to -\sigma$ under the time reversal.

The Kane-Mele model for the quantum spin Hall effect is a graphene model with the time reversal invariant spin-orbit coupling,

$$H = t \sum_{\langle i,j \rangle} c_i^\dagger c_j + i\lambda_{SO} \sum_{\langle\langle i,j \rangle\rangle} v_{ij} c_i^\dagger s_z c_j + i\lambda_R \sum_{\langle i,j \rangle} c_i^\dagger (s \times \mathbf{d}_{ij})_z \, c_j + \lambda_v \sum_i \xi_i c_i^\dagger c_i.$$

(6.13)

The first term is the nearest neighbor hopping term on a graphene lattice, where $c_i^\dagger = (c_{i,\uparrow}^\dagger, c_{i,\downarrow}^\dagger)$. The second term is a mirror symmetric spin-orbit interaction, which involves spin-dependent second neighbor hopping. Here $v_{ij} = \frac{2}{\sqrt{3}}(\mathbf{d}_i \times \mathbf{d}_j)_z = \pm 1$, where $\mathbf{d}_i$ and $\mathbf{d}_j$ are two unit vectors along the two bonds the electron traverses going from site j to i. The Pauli matrices s_i describe the electron spin. The third term is the nearest neighbor Rashba term, which explicitly violates the $z \to -z$ mirror symmetry. The last term is a staggered sublattice potential with $\xi_i = \pm 1$. Inclusion of the Rashba term makes the system more complicated since s_z is no longer conserved and the electrons with spin-up and spin-down are coupled together.

Following the method in the Haldane model, a four-band Hamiltonian can always be expressed in terms of the Dirac matrices

$$H(k) = \sum_{a=1}^{5} d_a(k)\Gamma^a + \sum_{a<b=1}^{5} d_{ab}(k)\Gamma^{ab}. \tag{6.14}$$

Here the five Dirac matrices

$$\Gamma^a = (\sigma_x \otimes s_0, \sigma_z \otimes s_0, \sigma_y \otimes s_x, \sigma_y \otimes s_y, \sigma_y \otimes s_z) \tag{6.15}$$

$(a = 1, 2, 3, 4, 5)$ where the Pauli matrices σ_i represent the sublattice indices and

$$\Gamma^{ab} = \frac{1}{2i}[\Gamma^a, \Gamma^b]. \tag{6.16}$$

In this representation, the time reversal operator is $\Theta = i(\sigma_0 \otimes s_y,)K$. The five Dirac matrices are even under time reversal,

$$\Theta\Gamma^a\Theta^{-1} = \Gamma^a, \tag{6.17}$$

while the ten commutators are odd,

$$\Theta\Gamma^{ab}\Theta^{-1} = -\Gamma^{ab}. \tag{6.18}$$

To have a time reversal invariant Hamiltonian, the coefficients should satisfy the relations,

$$d_\alpha(-k) = d_\alpha(k); \tag{6.19a}$$

$$d_{ab}(-k) = -d_{ab}(k). \tag{6.19b}$$

Thus, the coefficients in the Kane-Mele model are

$$d_1 = t\left(1 + 2\cos\frac{k_x}{2}\cos\frac{\sqrt{3}k_y}{2}\right); \tag{6.20a}$$

$$d_2 = \lambda_v; \tag{6.20b}$$

$$d_3 = \lambda_R\left(1 - \cos\frac{k_x}{2}\cos\frac{\sqrt{3}k_y}{2}\right); \tag{6.20c}$$

$$d_4 = -\sqrt{3}\lambda_R\sin\frac{k_x}{2}\sin\frac{\sqrt{3}k_y}{2}; \tag{6.20d}$$

$$d_{12} = -2t\cos\frac{k_x}{2}\sin\frac{\sqrt{3}k_y}{2}; \tag{6.20e}$$

$$d_{15} = \lambda_{SO} \left(2 \sin k_x - 4 \sin \frac{k_x}{2} \cos \frac{\sqrt{3}k_y}{2} \right);$$ (6.20f)

$$d_{23} = -\lambda_R \cos \frac{k_x}{2} \sin \frac{\sqrt{3}k_y}{2};$$ (6.20g)

$$d_{24} = \sqrt{3}\lambda_R \sin \frac{k_x}{2} \cos \frac{\sqrt{3}k_y}{2}.$$ (6.20h)

This equation gives four energy bands. When two lower energy bands are fully occupied, the system becomes insulating if an energy gap exists between two upper bands and two lower bands. As the whole system does not break the time reversal symmetry, the Chern number is always zero. For $\lambda_R = 0$, the Hamiltonian is split into two independent parts,

$$H = \sum_{s=\uparrow,\downarrow} H_s$$ (6.21)

where

$$H_s = t \sum_{\langle i,j \rangle} c_{i,s}^{\dagger} c_{j,s} + i s \lambda_{SO} \sum_{\langle\langle i,j \rangle\rangle} v_{ij} c_{i,s}^{\dagger} c_{j,s} + \lambda_v \sum_i \xi_i c_{i,s}^{\dagger} c_{i,s}.$$ (6.22)

In this case, there is an energy gap with magnitude $\left| 6\sqrt{3}\lambda_{SO} - 2\lambda_v \right|$. For $\lambda_v > 3\sqrt{3}\lambda_{SO}$ the gap is dominated by λ_v, while for $\lambda_v < 3\sqrt{3}\lambda_{SO}$ the gap is dominated by λ_{SO}. For each H_s, we can define a spin-dependent Chern number. For $\lambda_v > 3\sqrt{3}\lambda_{SO}$, the corresponding Chern number is zero for both $H_\uparrow$ and $H_\downarrow$. However, for $\lambda_v < 3\sqrt{3}\lambda_{SO}$ the corresponding Chern number becomes nonzero,

$$n_s = \text{sgn}(s\lambda_{SO}).$$ (6.23)

Although the total Chern number $n = n_+ + n_- = 0$, their difference $n_+ - n_- = 2$ or -2. Thus, for $\lambda_v < 3\sqrt{3}\lambda_{SO}$, it is a combination of two independent quantum Hall phases with different chirality, that is, the quantum spin Hall system [4].

For $\lambda_R \neq 0$, the electrons with spin-up and spin-down will mix together, and we cannot separate the whole system into two independent parts as in the case of $\lambda_R = 0$. In other words, we could not introduce a spin-dependent Chern number to describe this new phase. Instead, Kane and Mele introduced Z_2 invariant to describe it.

For a strip sample, we adopt the periodic boundary condition in the x-direction such that k_x is a good quantum number. Exact diagonalization gives the energy dispersion of the system as a function of $k_x a$. It is found that there are two distinct phases: (a) a pair of the bands connects the conduction and valence bands and (b) no band connects the two bands as plotted in Fig. 6.3. Since the system is insulating and there exists an energy gap in the bulk, the bands connecting the conduction and

a   **b**

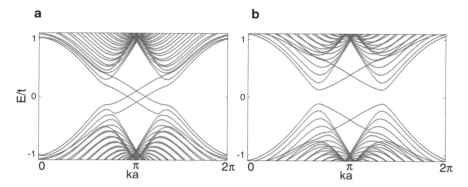

Fig. 6.3 Energy bands for a one-dimensional stripe with zigzag boundary condition. (**a**) Quantum spin Hall phases with $\lambda_v = 0.1t$ and (**b**) insulating phase with $\lambda_v = 0.4t$. In both cases, $\lambda_{SO} = 0.06t$ and $\lambda_R = 0.05t$

valence bands must be the edge states, which could be confirmed numerically. Thus, we conclude that in the topologically nontrivial phase, we have a pair of edge states between the bulk band gap at each boundary.

6.3 Transport of Edge States

The helical edge states are characteristic of two-dimensional topological insulator. It can be detected through the transport measurement in a mesoscopic device. Before we discuss the transport properties of the edge states in the quantum spin Hall system, we briefly introduce the Landauer-Büttiker formula for electron transport.

6.3.1 Landauer-Büttiker Formalism

Consider a one-dimensional conductor. Suppose the left side (the source) is filled up to the energy level μ_s, slightly higher than that of the right-hand side (the drain) μ_d. Then in the range between μ_s and μ_d, the conductor has been fully occupied states poring from left to right. The current through the channel is defined as

$$I = -ev_{\text{eff}}\delta N \tag{6.24}$$

where v_{eff} is the effective velocity of charge carriers along the channel near the Fermi energy and δN is the density of the charge carriers. Assume the voltage difference between two leads is quite small. Then

$$\delta N = D(E_f)(\mu_s - \mu_d) \tag{6.25}$$

where $D(E_f) = \partial N/\partial E|_{E_f}$ is the density of states at the Fermi level and $D(E_f) \approx D(\mu_s) \approx D(\mu_d)$. In one dimension, the velocity v_{eff} is given by the variance of dispersion with respect to the wave vector $v_{\text{eff}} = \partial E(k)/\hbar\partial k$ and the density of states $D(E) = \frac{\partial k/2\pi}{\partial E} = \frac{1}{h v_{\text{eff}}}$ and $\mu_s - \mu_d = -e(V_s - V_d)$. As a result, the current through the channel is given by

$$I = \frac{e^2}{h}(V_s - V_d). \tag{6.26}$$

The conductance is

$$G = \frac{I}{V_s - V_d} = \frac{e^2}{h}, \tag{6.27}$$

which is quantized in an ideal one-dimensional conductor.

More generally, Landauer proposed that the conductance of a mesoscopic conductor is given by [5,6],

$$G = \frac{2e^2}{h} M T, \tag{6.28}$$

where M stands for number of the transverse modes in the conductor and T is the averaged probability an electron injected from one end can transmit to the other. The factor 2 comes from the spin degeneracy of electron. The conductance is then independent of the system's dimension in length or width. Assume a conductor is connected to two electron reservoirs through ballistic leads. Then in the low temperature limit, the current flow is caused by the electrons' motion in the energy window $\mu_1 \sim \mu_2$. So the current transmitted from the left lead into the right lead is

$$I = -\frac{2e}{h} M T(\mu_1 - \mu_2), \tag{6.29}$$

and thus, the conductance is given by the linear response formula $G = I/\delta V$ ($\mu_1 - \mu_2 = -e\delta V$), which is exactly the Landauer formula in Eq. (6.28).

It can be shown that the Landauer formula recovers the classic Ohm's law in the large conductor scale limit. For a wide conductor, its number of conducting mode is proportional to the width W: $M \propto W$. Assume the conductor is long, we can prove that its transmission probability is given by

$$T(L) = \frac{L_0}{L + L_0}, \tag{6.30}$$

where L_0 is a characteristic length and L is the conductor's length.

Büttiker developed an approach to systematically treat the voltage and current probes in a multiple terminal device [7,8], which has helped interpreting numerous mesoscopic experimental results since the 1980s. The approach was to extend the two-terminal Landauer formula and to sum over all the probes. In a multiterminal device, the current in the ith terminal is given by

$$I_i = -\frac{e}{h} \sum_{j \neq i} [T_{ji} \mu_i - T_{ij} \mu_j], \tag{6.31}$$

where μ_i is the Fermi energy in the ith probe and is related to the voltage through $V_i = -\mu_i/e$ and T_{ij} can be seen as the product of the number of modes and the transmission probability from the jth probe to the ith probe. The summation is over all the probes apart from probe i. The above formula can be written in terms of the interterminal transmission coefficient T_{ij} as

$$I_i = \frac{e^2}{h} \sum_{j \neq i} [T_{ji} V_i - T_{ij} V_j]. \tag{6.32}$$

In the equilibrium condition, all the probes have equal voltage and zero current flow. And thus, from the above equation we must have

$$\sum_{j \neq i} T_{ji} = \sum_{j \neq i} T_{ij}, \tag{6.33}$$

which enables us to rewrite the Büttiker formula in Eq. (6.32) in a more straight-forward form:

$$I_i = \frac{e^2}{h} \sum_{j \neq i} T_{ij} [V_i - V_j]. \tag{6.34}$$

The Büttiker formula in Eq. (6.34) enables us to write the multiterminal conductance and resistance in compact forms of matrices. For example, without knowing the specific pattern of a three-terminal device, we know the current and voltage in the terminals are related to a set of equations:

$$\begin{pmatrix} I_1 \\ I_2 \\ I_3 \end{pmatrix} = \frac{e^2}{h} \begin{pmatrix} T_{12} + T_{13} & -T_{12} & -T_{13} \\ -T_{21} & T_{21} + T_{23} & -T_{23} \\ -T_{31} & -T_{32} & T_{31} + T_{32} \end{pmatrix} \begin{pmatrix} V_1 \\ V_2 \\ V_3 \end{pmatrix}. \tag{6.35}$$

This matrix equation can be further simplified by the fact that total current flow is conserved, that is, $I_1 + I_2 + I_3 = 0$. Also we know from the Landauer formula as well as the Büttiker formula that it is only the voltage difference between the probes that determines the magnitude of the current. Thus, we can set an arbitrary probe voltage to be 0. For instance, we can set $V_3 = 0$, and this enables us to reduce the matrix dimension by 1

$$\begin{pmatrix} I_1 \\ I_2 \end{pmatrix} = \frac{e^2}{h} \begin{pmatrix} T_{12} + T_{13} & -T_{12} \\ -T_{21} & T_{21} + T_{23} \end{pmatrix} \begin{pmatrix} V_1 \\ V_2 \end{pmatrix}. \tag{6.36}$$

The resistance is also in the matrix form, related to the conductance matrix through

$$\begin{pmatrix} R_{11} & R_{12} \\ R_{21} & R_{22} \end{pmatrix} = \frac{h}{e^2} \begin{pmatrix} T_{12} + T_{13} & -T_{12} \\ -T_{21} & T_{21} + T_{23} \end{pmatrix}^{-1}. \qquad (6.37)$$

The above approach has become a standard technique in finding the conductance and resistance in a multiterminal device.

6.3.2   Transport of Edge States

In the quantum spin Hall system, a pair of helical edge states consists two chiral states of electrons with spin-up ($\sigma = \uparrow$) and spin-down ($\sigma = \downarrow$). The transmission coefficient of the chiral state with spin-up from one terminal to its neighbor terminal, say, in the clockwise direction is $T_{ij}^{\uparrow} = 1$, and the transmission coefficient from one terminal to its neighbor terminal in the counterclockwise direction is $T_{ji}^{\uparrow} = 0$. Meanwhile, the transmission coefficient of the chiral state with spin-down from one terminal to its neighbor terminal in the clockwise direction is $T_{ij}^{\downarrow} = 0$, and the transmission coefficient from one terminal to its neighbor terminal in the counterclockwise direction is $T_{ji}^{\downarrow} = 1$.

The charge current at the terminal i is defined as the summation of the currents with spin-up and spin-down

$$I_i^c \equiv I_i^{\uparrow} + I_i^{\downarrow} = \frac{e^2}{h} \sum_{j \neq i, \sigma} (T_{ij}^{\sigma} V_j - T_{ji}^{\sigma} V_i). \qquad (6.38)$$

The spin current at the terminal i is defined as the difference of the currents with spin-up and spin-down

$$I_i^s \equiv \frac{\hbar/2}{e} \left(I_i^{\uparrow} - I_i^{\downarrow} \right) = \frac{e}{4\pi} \sum_{j \neq i, \sigma} \sigma (T_{ij}^{\sigma} V_j - T_{ji}^{\sigma} V_i) \qquad (6.39)$$

where we convert the unit of charge current into that of spin current: change the unit of charge e into the unit of spin $\hbar/2$ by the ratio $\frac{\hbar/2}{e}$.

Two-Terminal Measurement: As the edge states are helical, the transmission coefficients $T_{12}^{\sigma} = T_{21}^{\sigma} = 1$ for electrons with both spin-up ($\sigma = \uparrow$) and spin-down ($\sigma = \downarrow$). Take $V_1 = V/2$ and $V_2 = -V/2$. The spin-dependent current flowing out of the terminal 2 is

$$I_2^{\uparrow} = I_2^{\downarrow} = \frac{e^2}{h} (V_1 - V_2). \qquad (6.40)$$

Thus, the charge conductance is

$$G = G_\uparrow + G_\downarrow = 2\frac{e^2}{h} \tag{6.41}$$

since there are two conducting channels from the left to the right. This is equivalent to a quantum Hall conductance for $n = 2$ in a setup with two terminals.

Four-Terminal Measurement: In this case, the transmission coefficients for electron with spin-up $T_{43}^\uparrow = T_{32}^\uparrow = T_{21}^\uparrow = T_{14}^\uparrow = 1$ and 0 otherwise, and the transmission coefficients for electron with spin-down $T_{12}^\downarrow = T_{23}^\downarrow = T_{34}^\downarrow = T_{41}^\downarrow = 1$ and 0 otherwise. From the Landauer-Büttiker formula, we have

$$\begin{pmatrix} I_1^\uparrow \\ I_2^\uparrow \\ I_3^\uparrow \\ I_4^\uparrow \end{pmatrix} = \begin{pmatrix} -1 & 0 & 0 & 1 \\ 1 & -1 & 0 & 0 \\ 0 & 1 & -1 & 0 \\ 0 & 0 & 1 & -1 \end{pmatrix} \begin{pmatrix} V_1 \\ V_2 \\ V_3 \\ V_4 \end{pmatrix} \tag{6.42}$$

and

$$\begin{pmatrix} I_1^\downarrow \\ I_2^\downarrow \\ I_3^\downarrow \\ I_4^\downarrow \end{pmatrix} = \begin{pmatrix} -1 & 1 & 0 & 0 \\ 0 & -1 & 1 & 0 \\ 0 & 0 & -1 & 1 \\ 1 & 0 & 0 & -1 \end{pmatrix} \begin{pmatrix} V_1 \\ V_2 \\ V_3 \\ V_4 \end{pmatrix}. \tag{6.43}$$

The total charge current is the sum of the currents with spin-up and spin-down, $I_i = I_i^\uparrow + I_i^\downarrow$. Thus, the equation for the charge current

$$\begin{pmatrix} I_1 \\ I_2 \\ I_3 \\ I_4 \end{pmatrix} = \frac{e^2}{h} \begin{pmatrix} -2 & 1 & 0 & 1 \\ 1 & -2 & 1 & 0 \\ 0 & 1 & -2 & 1 \\ 1 & 0 & 1 & -2 \end{pmatrix} \begin{pmatrix} V_1 \\ V_2 \\ V_3 \\ V_4 \end{pmatrix}. \tag{6.44}$$

The total spin current in each terminal is the difference of the currents with spin-up and spin-down, $I_i^s = \left(I_i^\uparrow - I_i^\downarrow \right) \times \frac{\hbar/2}{e}$,

$$\begin{pmatrix} I_1^s \\ I_2^s \\ I_3^s \\ I_4^s \end{pmatrix} = \frac{e}{4\pi} \begin{pmatrix} 0 & -1 & 0 & 1 \\ 1 & 0 & -1 & 0 \\ 0 & 1 & 0 & -1 \\ -1 & 0 & 1 & 0 \end{pmatrix} \begin{pmatrix} V_1 \\ V_2 \\ V_3 \\ V_4 \end{pmatrix}. \tag{6.45}$$

Set the voltages at the terminals 1 and 3 $V/2$ and $-V/2$ and 0 for terminals 2 and 4

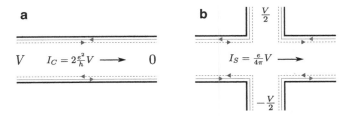

Fig. 6.4 Schematic diagram showing (**a**) two-terminal and (**b**) four-terminal measurement geometries. In (**a**), a charge current $I_C = (2e^2/h)V$ flows into the right lead. In (**b**), a spin current $I_S = \frac{e}{4\pi}V$ flows into the right lead

$$\begin{pmatrix} V_1 \\ V_2 \\ V_3 \\ V_4 \end{pmatrix} = \begin{pmatrix} \frac{V}{2} \\ 0 \\ -\frac{V}{2} \\ 0 \end{pmatrix}.$$ (6.46)

The currents at terminal 2 are

$$I_2^\uparrow = \frac{e^2}{h} T_{21}^\uparrow V_1 = +\frac{e^2}{2h} V,$$ (6.47a)

$$I_2^\downarrow = \frac{e^2}{h} T_{23}^\downarrow V_3 = -\frac{e^2}{2h} V.$$ (6.47b)

As a result, the total current is

$$I_2^c = I_2^\uparrow + I_2^\downarrow = 0.$$ (6.48)

However,

$$I_2^s = \left(I_2^\uparrow - I_2^\downarrow \right) \times \frac{\hbar/2}{e} = \frac{e}{4\pi} V.$$ (6.49)

Thus, the spin Hall conductance is $G_s = \frac{e}{4\pi}$. However, it is noted that the Hall conductance for each sector is

$$G^\uparrow = \frac{I_2^\uparrow}{V_1 - V_3} = \frac{e^2}{2h}$$ (6.50)

as we have set the voltages at the four-terminals. In conclusion, the quantum spin Hall effect can be measured through the charge transport in a mesoscopic system as shown in Fig. 6.4 [3].

Strictly speaking, the spin-up and spin-down here mean two different conducting channels of the edge states, not the real electron spin. Due to the spin-orbit coupling, none of the spin components S_α ($\alpha = x, y, z$) is conserved. So the "real" spin Hall conductance is not quantized.

6.4 Stability of Edge States

Assume the impurity potential V is time reversal invariant. There does not exist backscattering between the two helical edge states. The time reversal operator $\Theta^2 = -1$. Thus, we set $|u_{-k,\downarrow}\rangle = \Theta |u_{k,\uparrow}\rangle$ and $|u_{k,\uparrow}\rangle = -\Theta |u_{-k,\downarrow}\rangle$. The operator Θ is anti-unitary and has the property,

$$\langle \Theta\alpha | V | \Theta\beta \rangle = \langle \beta | V | \alpha \rangle. \tag{6.51}$$

Using this relation, it is easy to conclude

$$\langle u_{k,\uparrow} | V | u_{-k,\downarrow} \rangle = 0. \tag{6.52}$$

Li and Shi proposed a general argument for the robustness of the helical edge state transport [9]. In essence, a two-dimensional quantum spin Hall insulator is a conductor with an odd number of Kramers pairs of conducting channel. This is different from the ordinary one-dimensional conductor which always has an even number of Kramers pairs of conducting channel. In general, the transmission along the conductor can be characterized by a $2N \times 2N$ S matrix, which relates the incoming (ψ_{in}) and outcoming (ψ_{out}) wave amplitudes,

$$\psi_{\text{out}} = S\psi_{\text{in}} \tag{6.53}$$

where

$$\psi_{\text{in}} = (a_1^+, a_2^+, \cdots, a_N^+; b_1^-, b_2^-, \cdots, b_N^-)^T; \tag{6.54a}$$

$$\psi_{\text{out}} = (a_1^-, a_2^-, \cdots, a_N^-; b_1^+, b_2^+, \cdots, b_N^+)^T. \tag{6.54b}$$

$a_i^+ (b_i^+)$ and $a_i^- (b_i^-)$ denote the right-going and left-going wave amplitudes, respectively. $a_i^{\pm}$ and $b_i^{\pm}$ with the same index i are related by the time reversal and form a Kramers pair. N denotes the total number of Kramers pairs at each edge and can be odd for the quantum spin Hall insulator or even for ordinary insulator. In this notation, time reversal symmetry imposes the constraint on the S matrix

$$S^T = -S. \tag{6.55}$$

Moreover, the conservation of charge implies that the S matrix must be unitary: $S^\dagger S = 1$.

Under these constraints, the polar decomposition of the S matrix reads as [10]

$$S = \begin{pmatrix} U^T & 0 \\ 0 & V^T \end{pmatrix} \begin{pmatrix} \Sigma & T \\ -T & -\Sigma \end{pmatrix} \begin{pmatrix} U & 0 \\ 0 & V \end{pmatrix} \tag{6.56}$$

where U and V are two $N \times N$ unitary matrix. For even $N = 2n$, Σ is a block diagonal matrix

$$\Sigma = \sqrt{1 - T_1} i\sigma_y \oplus \sqrt{1 - T_2} i\sigma_y \oplus \cdots \oplus \sqrt{1 - T_n} i\sigma_y \tag{6.57}$$

and

$$T = \text{diag}[\sqrt{T_1}\sigma_0, \sqrt{T_2}\sigma_0, \cdots, \sqrt{T_n}\sigma_0]. \tag{6.58}$$

For odd $N = 2n + 1$, Σ is a block diagonal matrix

$$\Sigma = \sqrt{1 - T_1} i\sigma_y \oplus \sqrt{1 - T_2} i\sigma_y \oplus \cdots \oplus 0_{1\times1} \tag{6.59}$$

and

$$T = \text{diag}[\sqrt{T_1}\sigma_0, \sqrt{T_2}\sigma_0, \cdots, 1]. \tag{6.60}$$

T_i denotes the transmission coefficient in the ith conducting channel. One immediately sees that for odd N, there is at least one conducting channel that has the perfect transmission $T_i = 1$, without being adversely affected by the disorder. This is the reason behind the robustness of helical edge states in the quantum spin Hall effect.

According to the Z_2 classification for time reversal invariant insulating system, there always exist an odd number of Kramers pairs of conducting edge states along each edge of sample. However, in the geometry of a strip, there are two edges, and the total number of Kramers pairs is still even in the system. The conductance is not really quantized if the interaction or finite size effect makes the channels at two edges couple together [11].

6.5   Realization of Quantum Spin Hall Effect in HgTe/CdTe Quantum Well

In 2006 Bernevig, Hughes and Zhang [12] predicted that HgTe/CdTe quantum well may have an inverted band structure and exhibit the quantum spin Hall effect. One year later, König et al. [13] verified the theoretical prediction experimentally.

6.5.1   Band Structure of HgTe/CdTe Quantum Well

The band structures of HgTe and CdTe near the Γ point can be described very well by the six-band bulk Kane model which incorporates the Γ_6 and Γ_8 bands but neglects the split-off Γ_7 band. CdTe has a so-called normal band structure, in which the band Γ_6 of s-wave electron ($j = \frac{1}{2}$) has a higher energy and the band Γ_8 ($j = \frac{3}{2}$) has a lower energy. However, HgTe has an inverted band structure as shown in Fig. 6.5. In order to consider the coupling between the Γ_6 and Γ_8 bands, we choose the six-band basis set [14, 15],

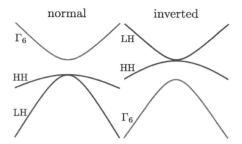

Fig. 6.5 A schematic illustration of the normal band structure and inverted band structure. The *left* is the normal band structure where the *blue curve* represents the light hole (LH) and heavy hole (HH) of the Γ_8 valence band, the *right* is the inverted band structure where the LH flips up and becomes the conduction band, and the Γ_6 appears below the HH band (Adapted from [15])

$$u_1(\mathbf{r}) = \left|\Gamma_6, +\frac{1}{2}\right\rangle_c = S\uparrow, \tag{6.61a}$$

$$u_2(\mathbf{r}) = \left|\Gamma_6, -\frac{1}{2}\right\rangle_c = S\downarrow, \tag{6.61b}$$

$$u_3(\mathbf{r}) = \left|\Gamma_8, +\frac{3}{2}\right\rangle_v = \frac{1}{\sqrt{2}}(X+iY)\uparrow, \tag{6.61c}$$

$$u_4(\mathbf{r}) = \left|\Gamma_8, +\frac{1}{2}\right\rangle_v = \frac{1}{\sqrt{6}}[(X+iY)\downarrow -2Z\uparrow], \tag{6.61d}$$

$$u_5(\mathbf{r}) = \left|\Gamma_8, -\frac{1}{2}\right\rangle_v = -\frac{1}{\sqrt{6}}[(X-iY)\uparrow +2Z\downarrow], \tag{6.61e}$$

$$u_6(\mathbf{r}) = \left|\Gamma_8, -\frac{3}{2}\right\rangle_v = -\frac{1}{\sqrt{2}}(X-iY)\downarrow. \tag{6.61f}$$

For the chosen basis set, the Hamiltonian for a three-dimensional system with [001] growth direction takes the following form:

$$H = \begin{pmatrix} T & 0 & -\frac{1}{\sqrt{2}}Pk_+ & \sqrt{\frac{2}{3}}Pk_z & \frac{1}{\sqrt{6}}Pk_- & 0 \\ 0 & T & 0 & -\frac{1}{\sqrt{6}}Pk_+ & \sqrt{\frac{2}{3}}Pk_z & \frac{1}{\sqrt{2}}Pk_- \\ -\frac{1}{\sqrt{2}}Pk_- & 0 & U+V & -S_- & R & 0 \\ \sqrt{\frac{2}{3}}Pk_z & -\frac{1}{\sqrt{6}}Pk_- & -S_-^\dagger & U-V & 0 & R \\ \frac{1}{\sqrt{6}}Pk_+ & \sqrt{\frac{2}{3}}Pk_z & R^\dagger & 0 & U-V & S_+^\dagger \\ 0 & \frac{1}{\sqrt{2}}Pk_+ & 0 & R^\dagger & S_+ & U+V \end{pmatrix} \tag{6.62}$$

where

$$k_{\parallel}^2 = k_x^2 + k_y^2, k_{\pm} = k_x \pm i k_y, \tag{6.63a}$$

$$T = E_c(z) + \frac{\hbar^2}{2m_0}\left[(2F+1)k_{\parallel}^2 + k_z(2F+1)k_z\right], \tag{6.63b}$$

$$U = E_v(z) - \frac{\hbar^2}{2m_0}\left(\gamma_1 k_{\parallel}^2 + k_z\gamma_1 k_z\right), \tag{6.63c}$$

$$V = -\frac{\hbar^2}{2m_0}\left(\gamma_2 k_{\parallel}^2 - 2k_z\gamma_2 k_z\right), \tag{6.63d}$$

$$R = -\frac{\hbar^2}{2m_0}\frac{\sqrt{3}}{2}\left[(\gamma_3 - \gamma_2)k_+^2 - (\gamma_3 + \gamma_2)k_-^2\right], \tag{6.63e}$$

$$S_{\pm} = -\frac{\hbar^2}{2m_0}\sqrt{3}k_{\pm}(\gamma_3 k_z + k_z\gamma_3). \tag{6.63f}$$

$P = -\frac{\hbar}{m_0}\langle s| p_x |X\rangle$ is the Kane matrix element between the s and p bands with m_0 the bare electron mass.

The quantum well growth direction is along z with $Hg_{1-x}Cd_xTe$ for $z < -d/2$, HgTe for $-d/2 < z < d/2$ and $Hg_{1-x}Cd_xTe$ for $z > d/2$. As the quantum well is confined along the z-direction, we make the substitution, $k_z = -i\frac{\partial}{\partial z}$. Now the problem reduces to solving, in the presence of continuous boundary conditions, the Hamiltonian (6.62) in each of the three regions of the quantum well.

The basic idea to derive an effective Hamiltonian is to start with the Hamiltonian at $k_x = k_y = 0$ and to find the solutions of the wave function of electrons in the confined quantum well. Then using the solution of $k_x = k_y = 0$ as the basis, one can derive an effective Hamiltonian for $k_x, k_y \neq 0$ by means of the projected perturbation method.

For $k_x = k_y = 0$,

$$H(k_{\parallel} = 0) = \begin{pmatrix} T & 0 & 0 & \sqrt{\frac{2}{3}}Pk_z & 0 & 0 \\ 0 & T & 0 & 0 & \sqrt{\frac{2}{3}}Pk_z & 0 \\ 0 & 0 & U+V & 0 & 0 & 0 \\ \sqrt{\frac{2}{3}}Pk_z & 0 & 0 & U-V & 0 & 0 \\ 0 & \sqrt{\frac{2}{3}}Pk_z & 0 & 0 & U-V & 0 \\ 0 & 0 & 0 & 0 & 0 & U+V \end{pmatrix}, \tag{6.64}$$

which is reduced to a block diagonalized matrix after rearranging the basis as $\{u_1, u_4, u_3, u_2, u_5, u_6\}$. On the subsector of $\{u_1, u_4\}$ for $j_z = \frac{1}{2}$,

$$H_{\text{eff}} = \begin{pmatrix} T & \sqrt{\frac{2}{3}}Pk_z \\ \sqrt{\frac{2}{3}}Pk_z & U-V \end{pmatrix}, \tag{6.65}$$

which is a one-dimensional modified Dirac equation. Consider a quantum well potential $V_{QW}(z)$. The model coefficients are different for CdTe at $|z| > d/2$ and HgTe at $|z| < d/2$. Solving this one-dimensional problem, one obtains a bound state for the quantum well φ_1. Similarly, on the base $\{u_3\}$ of $j_z = \frac{3}{2}$, one obtains a solution for quantum well φ_2. Using these two states, one can have an effective Hamiltonian near the point of $k \neq 0$,

$$h(k) = (\langle \varphi_1 | , \langle \varphi_2 |) H(k) \begin{pmatrix} |\varphi_1\rangle \\ |\varphi_2\rangle \end{pmatrix}. \tag{6.66}$$

(u_2, u_5, u_6) gives other two states. In this way, Bernevig, Hughes, and Zhang derived an effective model for a quantum well of HgTe/CdTe [12],

$$H_{\mathrm{BHZ}} = \begin{pmatrix} h(k) & 0 \\ 0 & h^*(-k) \end{pmatrix} \tag{6.67}$$

where $h(k) = \epsilon(k) + A(k_x \sigma_x + k_y \sigma_y) + (M - Bk^2)\sigma_z$.

The model is actually equivalent to the modified two-dimensional Dirac model as shown in Eq. (2.42) in addition of the kinetic term $\epsilon(k)$,

$$h(k) = \epsilon(k) + h_+, \tag{6.68}$$

$$h^*(-k) = \epsilon(k) + Uh_-U^{-1}, \tag{6.69}$$

where the unitary transformation matrix $U = \sigma_z$. All the model parameters are functions of the thickness of quantum well. The most striking property of this system is that the mass or gap parameter M changes sign when the thickness d of the quantum well is varied through a critical thickness d_c ($= 6.3$ nm) associating with the transition of electronic band structure from a normal to an "inverted" type [16].

If the inclusion of $\epsilon(k)$ does not close the energy gap caused by M for a nonzero B, the system should be insulating in the bulk. There exists a topological phase transition from positive M to negative M. However, the sign of M alone cannot determine whether the system is topologically trivial or nontrivial. From the formula in Eq. (2.48), we know that the system is in a quantum spin Hall phase only for $MB > 0$ and there exists a pair of helical edge states at the boundary of the system.

6.5.2 Exact Solution of Edge States

In this subsection, we present an exact solution of edge state for the Bernevig-Hughes-Zhang model in Eq. (6.67), which was first solved in the paper by Zhou et al. [11]. Here we consider a semi-infinite plane with an open boundary condition at $y = 0$. In this case, k_x is a good quantum number, and k_y is replaced by using the substitution $k_y = -i\partial_y$. The Hamiltonian is a block-diagonalized one,

$$\mathcal{H}\left(k_x, -i\,\partial_y\right) = \begin{pmatrix} h_\uparrow\left(k_x, -i\,\partial_y\right) & 0 \\ 0 & h_\downarrow\left(k_x, -i\,\partial_y\right) \end{pmatrix}, \tag{6.70}$$

where

$$h_\uparrow\left(k_x, -i\,\partial_y\right) = \begin{pmatrix} M - B_+\left(k_x^2 - \partial_y^2\right) & A(k_x - \partial_y) \\ A(k_x + \partial_y) & -M + B_-\left(k_x^2 - \partial_y^2\right) \end{pmatrix} \tag{6.71}$$

and

$$h_\downarrow\left(k_x, -i\,\partial_y\right) = \begin{pmatrix} M - B_+\left(k_x^2 - \partial_y^2\right) & -A(k_x + \partial_y) \\ -A(k_x - \partial_y) & -M + B_-\left(k_x^2 - \partial_y^2\right) \end{pmatrix}. \tag{6.72}$$

with $B_\pm = B \pm D$. The upper $h_\uparrow$ and lower $h_\downarrow$ blocks describe the states of spin-up (strictly speaking, it is the sector of $j_z = \frac{1}{2}$ and $\frac{3}{2}$) and spin-down (the sector of $j_z = -\frac{1}{2}$ and $-\frac{3}{2}$), respectively.

The eigenvalue problem of the upper and lower blocks can be solved separately. Here we focus on the solution for the upper block of this Hamiltonian,

$$h_\uparrow \Psi_\uparrow = E \Psi_\uparrow. \tag{6.73}$$

We take the trial wave function

$$\Psi_\uparrow = \begin{pmatrix} c \\ d \end{pmatrix} e^{\lambda y} \tag{6.74}$$

and substitute it into Eq. (6.73). Then the characteristic equation gives

$$\det \begin{pmatrix} M - B_+\left(k_x^2 - \lambda^2\right) - E & A(k_x - \lambda) \\ A(k_x + \lambda) & -M + B_-\left(k_x^2 - \lambda^2\right) - E \end{pmatrix} = 0. \tag{6.75}$$

We obtain four real roots $\pm\lambda_1$ and $\pm\lambda_2$,

$$\lambda_{1,2}^2 = k_x^2 + F \pm \sqrt{F^2 - \frac{M^2 - E^2}{B_+ B_-}}, \tag{6.76}$$

where $F = [A^2 - 2(MB + ED)]/(2B_+B_-)$. To find an edge state solution, the wave function must decay to zero when deviating far away from the boundary. We adopt the Dirichlet boundary condition $\Psi_\uparrow(k_x, y = 0) = \Psi_\uparrow(k_x, y = +\infty) = 0$. Thus, the solution has a general form:

$$\Psi_\uparrow = \begin{pmatrix} \tilde{c}(k_x) \\ \tilde{d}(k_x) \end{pmatrix} (e^{-\lambda_1 y} - e^{-\lambda_2 y}), \tag{6.77}$$

if λ_1 and λ_2 are positive or their real parts are positive. Substituting the solution into Eq. (6.73), we obtain

$$\frac{\tilde{c}}{\tilde{d}} = \frac{A(k_x + \lambda_1)}{E - M + B_+ k_x^2 - B_+ \lambda_1^2} = \frac{A(k_x + \lambda_2)}{E - M + B_+ k_x^2 - B_+ \lambda_2^2}. \tag{6.78}$$

Thus, it follows from this equation that

$$E = M - B_+ \lambda_1 \lambda_2 - B_+ (\lambda_1 + \lambda_2) k_x - B_+ k_x^2. \tag{6.79}$$

At $k_x = 0$,

$$E = M - B_+ \lambda_1 \lambda_2, \tag{6.80a}$$

$$\lambda_1 = \sqrt{F + \sqrt{F^2 - (M^2 - E^2)/B_+ B_-}}, \tag{6.80b}$$

$$\lambda_2 = \sqrt{F - \sqrt{F^2 - (M^2 - E^2)/B_+ B_-}}. \tag{6.80c}$$

Thus, one obtains

$$E = \frac{B_- - B_+}{B_- + B_+} M = -\frac{D}{B} M, \tag{6.81}$$

$$\lambda_1 \lambda_2 = \frac{M - E}{B_+} = \frac{M}{B} > 0, \tag{6.82}$$

$$\lambda_1 + \lambda_2 = \sqrt{\frac{A^2}{B_+ B_-}} > 0. \tag{6.83}$$

Therefore, the existing conditions of the edge state solution are

$$\frac{A^2}{B_+ B_-} > 0, \frac{M}{B} > 0. \tag{6.84}$$

Near $k_x = 0$, from the equations for λ_1, λ_2, and E, we calculate

$$\left. \frac{dE}{dk_x} \right|_{k_x=0} = -B_+ \left. \frac{d(\lambda_1 \lambda_2)}{dk_x} \right|_{k_x=0} - B_+ (\lambda_1 + \lambda_2)|_{k_x=0}$$

$$= \text{sgn}(B) A \sqrt{1 - \frac{D^2}{B^2}}. \tag{6.85}$$

It follows that the energy spectrum of the edge states near $k_x = 0$ reads

$$E_\uparrow (k_x) = -\frac{MD}{B} + \text{sgn}(B) A \sqrt{1 - \frac{D^2}{B^2}} k_x + O(k_x^2). \tag{6.86}$$

The effective velocity of this state is

$$v_\uparrow = +\text{sgn}(B)A\sqrt{1 - \frac{D^2}{B^2}}. \tag{6.87}$$

Similarly, we may have the energy dispersion of the edge states for the lower block $h_\downarrow$

$$E_\downarrow(k_x) = -\frac{MD}{B} - \text{sgn}(B)A\sqrt{1 - \frac{D^2}{B^2}}k_x + O(k_x^2) \tag{6.88}$$

and the effective velocity

$$v_\downarrow = -\text{sgn}(B)A\sqrt{1 - \frac{D^2}{B^2}}. \tag{6.89}$$

The results can also be obtained from the perturbation theory for a small k_x. Thus, the effective velocities in two edge states are opposite; one is positive and the other is negative. These two edge states constitute a pair of helical edge states.

6.5.3   Experimental Measurement

The transition from the normal band to an inverted band structure coincides with the topological quantum phase transition from a trivial insulator to a quantum spin Hall insulator. The first experimental observation was made by a group in Wurzburg, Germany, led by Laurens W. Molenkamp [13]. In order to cover the regime of normal and inverted band structure, a series of HgTe samples with the quantum well width from 4.5 to 12 nm were grown. Initial evidence for the quantum spin Hall state was revealed when the Hall bar of dimension $(L \times W) = (20.0 \times 13.3)\,\mu\text{m}^2$ with different thickness were studied. For a thinner sample with $d_{QW} < d_c$, the sample shows an insulating behavior. But for a thicker sample with $d_{QW} > d_c$, a finite value of resistance was measured as shown in Fig. 6.6, which is anticipated as the theoretical prediction for the quantum spin Hall effect. The inset shows the resistances at two different temperatures. Surprisingly, the resistance at lower temperature is larger than one at higher temperature, which usually is characteristic of an insulating phase rather than a conducting phase. We have to say that no quantized conductance has been yet measured experimentally in quantum spin Hall effect, although the measured conductance is close to the predicted value at a specific temperature.

Further evidence for the helical edge state comes from the nonlocal transport measurement, which is performed in a multiterminal setup. In conventional diffusive electronics, bulk transport satisfies Ohm's law. The resistance is proportional to the length and inversely proportional to the cross-sectional area, implying the existence of a local resistivity or conductivity tensor. However, the existence of the edge state

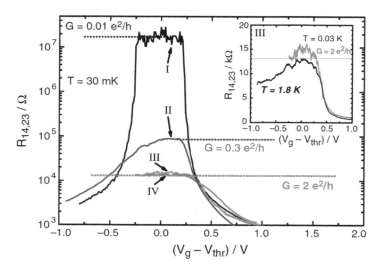

Fig. 6.6 The longitudinal four-terminal resistance, $R_{14,23}$, of various normal ($d = 5.5\,\text{nm}$) (I) and inverted ($d = 7.3\,\text{nm}$) (II, III, and IV) quantum well structures as a function of the gate voltage measured for $B = 0\,\text{T}$ at $T = 30\,\text{mK}$. The device sizes are $(20.0 \times 13.3)\,\mu\text{m}^2$ for devices I and II, $(1.0 \times 1.0)\,\mu\text{m}^2$ for device III, and $(1.0 \times 0.5)\,\mu\text{m}^2$ for device IV. The inset shows $R_{14,23}(V_g)$ of two samples from the same wafer, having the same device size (III) at 30 mK (*green*) and 1.8 K (*black*) on a linear scale (Adapted from [13]. Reprinted with permission from AAAS)

Fig. 6.7 An H-shaped four-terminal device of the quantum spin Hall system. The spin-filtered edge channels are indicated by *red/solid* (spin-up) and *blue/dashed* (spin-down) *arrowed lines*

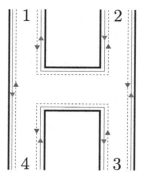

necessarily leads to nonlocal transport, which invalidates Ohm's law. Such nonlocal transport was first observed in the quantum Hall effect and is well described by the quantum transport theory based on the Landauer-Büttiker formula.

In the device shown in Fig. 6.7, which is used in the nonlocal measurements to prove the existence of helical edge states, two terminals act as current probes, and the other two act as voltage probes. The nonlocal resistance is defined as

$$R_{ij,kl} = \frac{V_k - V_l}{I_{ij}}. \qquad (6.90)$$

We can set $V_4 = 0$ and write down the Büttiker formula

$$\begin{pmatrix} I_1 \\ I_2 \\ I_3 \end{pmatrix} = \frac{e^2}{h} \begin{pmatrix} 2 & -1 & 0 \\ -1 & 2 & -1 \\ 0 & -1 & 2 \end{pmatrix} \begin{pmatrix} V_1 \\ V_2 \\ V_3 \end{pmatrix}. \tag{6.91}$$

If the current is driven through the terminals 1 and 4, and the terminals 2 and 3 act as voltage probes, we have $I_1 = -I_4$ and $I_2 = I_3 = 0$. Then we can solve this equation and get

$$R_{14,14} = \frac{V_1 - V_4}{I_1} = \frac{h}{e^2} \frac{3}{4}, \tag{6.92}$$

$$R_{14,23} = \frac{V_2 - V_3}{I_1} = \frac{h}{e^2} \frac{1}{4}, \tag{6.93}$$

which are the predicted value if helical edge states truly exists in the system. These predictions have been experimentally confirmed in HgTe/CdTe quantum well [17].

6.6 Quantum Hall Effect and Quantum Spin Hall Effect: A Case Study

The difference between the quantum Hall effect and the quantum spin Hall effect can be illustrated from the conductance of a three-probe conductor with one contact playing the role of a voltage probe. At such a contact, the net charge current vanishes. Electrons that leave the contact are replaced by the electrons from the contact reservoir. In the quantum Hall effect sample with $\nu = 2$, two edge states from the left source contact enter the voltage probe and two edge states leave the voltage probe to the right drain contact. The potential of the probe is equal to that of the source contact, and the voltage probe has no effect on the overall conductance. However, in the quantum spin Hall effect sample, the situation is different. Here, only one edge state is directed from the source contact to the voltage probe. Two other edge states lead away from the probe – one to the source contact and one to the sink contact. To maintain zero current, it is sufficient to tune the chemical potential at the probe halfway between the potentials of the source and drain contacts. Now, half the current is directed back to the source contact. The voltage probe reduces the overall conductance by half a conductance quantum, that is, $\sigma = \frac{3}{2} \frac{e^2}{h}$ not $2 \frac{e^2}{h}$ as in the quantum Hall effect of $\nu = 2$ [18]. Such a probe maintains zero net charge current into the contact. However, the spin current into the probe is nonzero and net spin-up in the case depicted. Simultaneously, a spin current is induced into both the source and drain electrodes.

Quantum Hall effect ($\nu = 2$): In this setup of three-terminal geometry, using the Landauer-Büttiker formula, the currents are

$$\begin{pmatrix} I_{\text{left}} \\ I_{\text{probe}} \\ I_{\text{right}} \end{pmatrix} = 2\frac{e^2}{h} \begin{pmatrix} -1 & 0 & 1 \\ 1 & -1 & 0 \\ 0 & 1 & -1 \end{pmatrix} \begin{pmatrix} \mu_{\text{left}} \\ \mu_{\text{probe}} \\ \mu_{\text{right}} \end{pmatrix}. \tag{6.94}$$

The probe potential is tuned such that the charge current at the voltage probe vanishes:

$$I_{\text{probe}} = 2\frac{e^2}{h}(\mu_{\text{left}} - \mu_{\text{probe}}) = 0. \tag{6.95}$$

Then

$$\begin{aligned} I_{\text{right}} &= 2\frac{e^2}{h}(\mu_{\text{probe}} - \mu_{\text{right}}) \\ &= 2\frac{e^2}{h}(\mu_{\text{left}} - \mu_{\text{right}}). \end{aligned} \tag{6.96}$$

The conductance is

$$G_{\text{QHE}} = 2\frac{e^2}{h}. \tag{6.97}$$

Quantum spin Hall effect: Using the Landauer-Büttiker formula, the currents are

$$\begin{pmatrix} I_{\text{left}} \\ I_{\text{probe}} \\ I_{\text{right}} \end{pmatrix} = \frac{e^2}{h} \begin{pmatrix} -2 & 1 & 1 \\ 1 & -2 & 1 \\ 1 & 1 & -2 \end{pmatrix} \begin{pmatrix} \mu_{\text{left}} \\ \mu_{\text{probe}} \\ \mu_{\text{right}} \end{pmatrix}. \tag{6.98}$$

The probe potential is tuned such that the charge current at the probe vanishes:

$$I_{\text{probe}} = \frac{e^2}{h}(\mu_{\text{left}} + \mu_{\text{right}} - 2\mu_{\text{probe}}) = 0. \tag{6.99}$$

Then

$$\begin{aligned} I_{\text{right}} &= \frac{e^2}{h}(\mu_{\text{left}} + \mu_{\text{probe}} - 2\mu_{\text{right}}) \\ &= \frac{3}{2}\frac{e^2}{h}(\mu_{\text{left}} - \mu_{\text{right}}). \end{aligned} \tag{6.100}$$

The charge conductance is

$$G_{\text{QHE}} = \frac{3}{2}\frac{e^2}{h}. \tag{6.101}$$

In the quantum spin Hall effect, the spin currents are

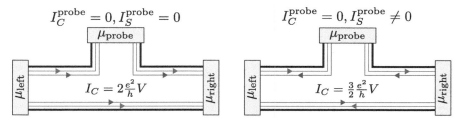

Fig. 6.8 Difference between the quantum Hall effect and quantum spin Hall effect (*right*) in a setup with three probes. *Left*: the quantum Hall effect for $\nu = 2$ with two chiral edge channels. *Right*: the quantum spin Hall effect with a pair of helical edge channels

$$\begin{pmatrix} I_{\text{left}}^S \\ I_{\text{probe}}^S \\ I_{\text{right}}^S \end{pmatrix} = \frac{e}{4\pi} \begin{pmatrix} 0 & -1 & 1 \\ 1 & 0 & -1 \\ -1 & 1 & 0 \end{pmatrix} \begin{pmatrix} \mu_{\text{left}} \\ \mu_{\text{probe}} \\ \mu_{\text{right}} \end{pmatrix}. \tag{6.102}$$

The spin current at the probe is

$$I_{\text{probe}}^S = \frac{e}{4\pi} \left(\mu_{\text{left}} - \mu_{\text{right}} \right) \neq 0. \tag{6.103}$$

Although the charge current vanishes at the probe, the spin current does not vanish. The results are summarized in Fig. 6.8.

6.7   Coherent Oscillation Due to the Edge States

We study here the device shown in Fig. 6.9, which consists of a two-dimensional strip of a topological insulator on which two quantum point contacts have been patterned in series through gates (shaded regions in Fig. 6.9). The quantum point contacts define a saddle-shaped confining potential, whose height can be controlled by a gate voltage. An effective disk of area $A = \pi R^2$ (R is the radius of the disk) is formed in the center. The Aharonov-Bohm effect in the device can be expected intuitively because a topological insulator possesses a pair of independent gapless edge states of different spins moving in opposite directions, each forming an ideal one-dimensional loop around the disk. The two edge states are independent because no backscattering is allowed at a given sample edge even in the presence of weak time reversal invariant disorder. We note here that spin is not a good quantum number in topological insulators because of spin-orbit coupling. In the absence of a magnetic field, the actual edge states are eigenstates of the time reversal operator; their characterization as spin-up and spin-down is not precisely correct, and the word "spin" below is to be viewed more generally as the quantum number denoting the two states of a Kramers doublet.

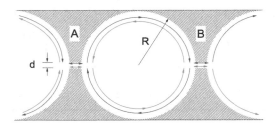

Fig. 6.9 Schematic of the setup consisting of a disk connected to two reservoirs through two quantum point contacts. *Red* (light gray) and *blue* (*dark gray*) *lines* indicate the chiral edge channels of spin-up and spin-down electrons, with *arrows* indicating the direction of their motion (Adapted from [20])

Suppose a weak magnetic field $B_\perp$ exists normal to the plane. Following Ref. [19], we consider a spin-up (or spin-down) electron travelling from the left-hand side in Fig. 6.9. At the left-hand side junction, it splits into two partial waves: one is transmitted through the quantum point contact into the disk with amplitude t, and the other is transmitted across the quantum point contact with an amplitude r causing a backscattering. We denote the wave function amplitudes in the upper and lower edges, right after the left-hand side junction, by u_1 and d_1, respectively. The corresponding amplitudes in the vicinity of right-hand side junction are $u_2 = u_1 \exp[i\theta/2]$ and $d_2 = d_1 \exp[-i\theta/2]$, where

$$\theta = 2\pi \frac{\phi}{\phi_0} + 2\pi k R_{\text{eff}}, \tag{6.104}$$

$\phi_0 = h/e$ is the magnetic flux quantum, $\phi = \pi R_{\text{eff}}^2 B_\perp$ is the magnetic flux threading the effective one-dimensional loop with an effective radius R_{eff}, and $2\pi k R_{\text{eff}}$ is the phase acquired by the wave function traveling along the loop. Since the Fermi level of electrons in the edge states can be tuned by a gate voltage V_{gate} and the dispersion relation for the edge states is linear in k, the phase can be tuned by the gate voltage $\delta\theta = 2\pi R_{\text{eff}}\delta k \propto \delta V_{\text{gate}}$.

A partial wave goes through the right-hand side slit with an amplitude t' and across the slit with an amplitude r'. Using the theory of multi-scattering processes [21], it follows that the total transmission for spin-up electron through the slit A and B is given by

$$T^\uparrow(B_\perp) = \frac{|tt'|^2}{1 + |rr'|^2 - 2|rr'|\cos(\theta + \theta_0)}. \tag{6.105}$$

Here $\theta_0 = \arg(rr')$. For specificity, our numerical calculations below suppose two symmetric quantum point contacts with $|t| = |t'|$ and $|r| = |r'|$. Resonant tunneling, that is, $T^\uparrow(B_\perp) = 1$, occurs for $\cos(\theta + \theta_0) = 1$.

The transmission coefficient for a spin-down electron $T^{\downarrow}(B_{\perp})$, which is the time reversal counterpart of spin-up electron at $-B_{\perp}$ field, is given by $T^{\downarrow}(B_{\perp}) = T^{\uparrow}(-B_{\perp})$. According to the Landauer-Büttiker formula [6, 22], the total conductance is

$$G(B) = \frac{e^2}{h}[T^{\uparrow}(B_{\perp}) + T^{\downarrow}(B_{\perp})]. \tag{6.106}$$

This coherence oscillations in the conductance G as a function of the magnetic flux ϕ through the disk are therefore expected to be symmetric with respect to the direction of the magnetic field.

6.8 Further Reading

- F.D.M. Haldane, Model for a quantum Hall effect without Landau levels: condensed-matter realization of the "Parity Anomaly". Phys. Rev. Lett. **61**, 2015 (1988)
- C.L. Kane, E.J. Mele, Quantum spin Hall effect in Graphene. Phys. Rev. Lett. **95**, 226801 (2005)
- C.L. Kane, E.J. Mele, Z_2 Topological order and the quantum spin Hall effect. Phys. Rev. Lett. **95**, 146802 (2005)
- B. Andrei Bernevig, T.L. Hughes, S.C. Zhang, Quantum spin Hall effect and topological phase transition in HgTe quantum wells. Science **314**, 1757 (2006)
- M. Konig, S. Wiedmann, Ch. Brüne, A. Roth, H. Buhmann, L.W. Molenkamp, X.L. Qi, S.C. Zhang, Quantum spin Hall insulator state in HgTe quantum wells. Science **318**, 766 (2007)

References

1. F.D.M. Haldane, Phys. Rev. Lett. **61**, 2015 (1988)
2. P. Strěda, J. Phys. C **15**, L717 (1982)
3. C.L. Kane, E.J. Mele, Phys. Rev. Lett. **95**, 226801 (2005)
4. D.N. Sheng, Z.Y. Weng, L. Sheng, F.D.M. Haldane, Phys. Rev. Lett. **97**, 036808 (2006)
5. R. Landauer, IBM J. Res. Dev. **1**, 223 (1957)
6. R. Landauer, Philos. Mag. **21**, 863 (1970)
7. M.Büttiker, Phys. Rev. Lett. **57**, 1761 (1986)
8. M.Büttiker, IBM J. Res. Dev. **32**, 317 (1988)
9. D.F. Li and J.R. Shi, Phys. Rev. B **79**, 241303(R) (2009)
10. C.W.J. Beenakker, Rev. Mod. Phys. **69**, 731 (1997)
11. B. Zhou, H.Z. Lu, R.L. Chu, S.Q. Shen, Q. Niu, Phys. Rev. Lett. **101**, 246807 (2008)
12. B.A. Bernevig, T.L. Hughes, S.C. Zhang, Science **314**, 1757 (2006)
13. M. König, S. Wiedmann, C. Brüne, A. Roth, H. Buhmann, L.W. Molenkamp, X.L. Qi, S.C. Zhang, Science **318**, 766 (2007)
14. E.G. Novik, A. Pfeuffer-Jeschke, T. Jungwirth, V. Latussek, C.R. Becker, G. Landwehr, H. Buhmann, L.W. Molenkamp, Phys. Rev. B **72**, 035321 (2005)

15. Chu, R.L., W.Y. Shan, J.Lu, S.Q. Shen, Phys. Rev. B **83**, 075110 (2011)
16. A. Pfeuffer-Jeschke, Ph.D. thesis, University of Wurzburg, 2000
17. A. Roth, C. Brüne, H. Buhmann, L.W. Molenkamp, J. Maciejko, X.L. Qi, S.C. Zhang, Science **325**, 294 (2009)
18. M. Büttiker, Science **325**, 278 (2009)
19. U. Sivan, Y. Imry, C. Hartzstein, Phys. Rev. B **39**, 1242 (1989)
20. R.L. Chu, W.Y. Shan, J. Lu, S.Q. Shen, Phys. Rev. B **83**, 075110 (2011)
21. S. Datta, *Electronic Transport in Mesoscopic Systems* (Cambridge University Press, Cambridge, 1995)
22. M. Büttiker, Phys. Rev. B **38**, 9375 (1988)

Chapter 7
Three-Dimensional Topological Insulators

Abstract Three-dimensional topological insulator is characterized by the surrounding surface states, in which electrons are well described as two-dimensional Dirac fermions. A series of materials have been discovered to be topological insulators since theoretical predictions.

Keywords Three-dimensional topological insulator • Surface states • Dirac cone • Surface quantum Hall effect • Topological insulator thin film

7.1 Family Members of Three-Dimensional Topological Insulators

7.1.1 Weak Topological Insulators: $Pb_x Sn_{1-x} Te$

The first known inverted band material is SnTe, which was discovered more than 50 years ago [1]. The valence- and conduction-band edges in PbTe and SnTe occur at the L points in the Brillouin zone. It was believed that the valence band of PbTe is an L_6^+ state and its conduction band is an L_6^- state while the valence band of SbTe is an L_6^- state and its conduction band is an L_6^+ state as shown in Fig. 7.1. In a $Pb_x Sn_{1-x}$ Te alloy sample, with increasing Sn composition, the energy gap initially decreases as the L_6^+ and L_6^- states approach each other, then closes at an intermediate composition where the two states become degenerate, and finally reopens, with the L_6^+ state now forming the conduction band and the L_6^- state forming the valence band [2]. The band structures of PbTe and SnTe were calculated in the early 1960s [3, 4]. It was realized that the change in the energy gap with composition for the $Pb_x Sn_{1-x}$ Te alloy series can be understood qualitatively in terms of the difference between the relativistic effect in Pb and Sn,

S.-Q. Shen, *Topological Insulators: Dirac Equation in Condensed Matters*,   113
Springer Series in Solid-State Sciences 174, DOI 10.1007/978-3-642-32858-9_7,
© Springer-Verlag Berlin Heidelberg 2012

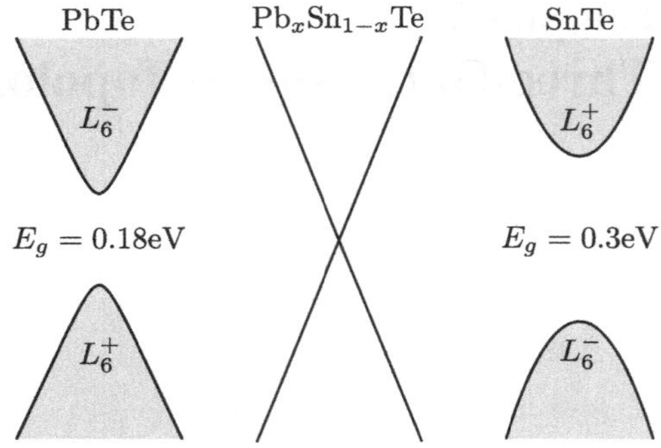

Fig. 7.1 Schmatic representation of the valence and conduction bands for PbTe, for the composition at which the energy gap is zero and for SnTe

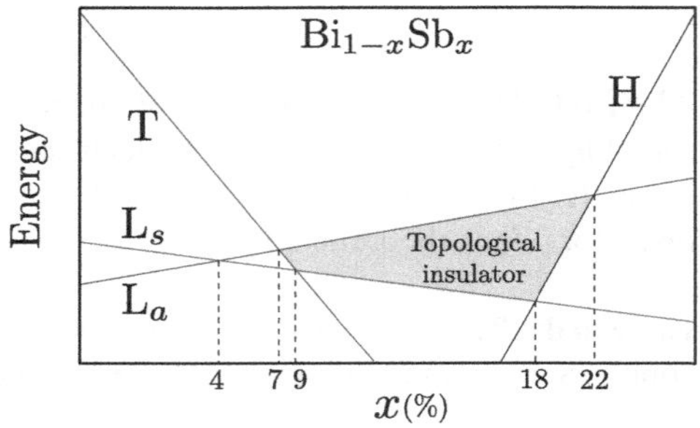

Fig. 7.2 Schematic representation of band energy evolution of $Bi_{1-x}Sb_x$ as a function of x (Replotted from [9])

and the relativistic correction is extremely important in determining the positions of the energy bands. Nowadays, we call the relativistic correction as the spin-orbit coupling in semiconductors.

The band inversion in $Pb_xSn_{1-x}Te$ occurs at four equivalent valleys. The number of the surface states is even. Thus it is a trivial or weak topological insulator according to the topological classification of time reversal symmetry [5].

7.1.2 Strong Topological Insulators: $Bi_{1-x}Sb_x$

The first discovered strong topological insulator is bismuth antimony alloy $Bi_{1-x}Sb_x$ [6,7]. Semiconducting $Bi_{1-x}Sb_x$ alloys have been studied experimentally because of their thermoelectric properties, which make them desirable for applications as thermocouple. The evolution of the band structure of the alloy $Bi_{1-x}Sb_x$ as a function of Sb composition x has been well studied and is summarized in Fig. 7.2 [8,9]. As the Sb concentration increases, two things happen. First, the gap between the L_s and L_a bands decreases. At $x = 4\%$, the band gap closes and then reopens with the inverted ordering. Second, the top of the valence band at T comes down in energy and crosses the bottom of the conduction band at $x = 7\%$. At this point, the indirect gap becomes positive, and the alloy is a semiconductor. At $x = 9\%$, the T valence band crosses the L_s valence band, and the alloy is a direct-gap semiconductor at the L points. As x increases further, the gap increases and reaches its maximum value of about 30 meV at $x = 18\%$. At that point, the valence band H crosses the L_s valence band. For $x > 22\%$, the H band crosses the L_a conduction band, and the alloy is again a semimetal. Since the inversion transition between the L_s and L_a bands occurs in the semimetal phase adjacent to pure bismuth, it is clear that the semiconducting $Bi_{1-x}Sb_x$ alloy inherits its topological class from pure antimony and is thus a strong topological insulator [6].

Direct observation of Dirac gapless surface states in $Bi_{1-x}Sb_x$ was first reported by a group led by Hasan [7]. High-momentum-resolution angle-resolved photoemission spectroscopy performed with varying incident photon energy allows for the measurement of electronic band dispersion along various momentum space (k-space) trajectories in the three-dimensional bulk Brillouin zone. The surface band-dispersion image along $\Gamma - M$ direction shows five Fermi level crossings, which indicates that these surface states are topologically nontrivial.

7.1.3   Topological Insulators with a Single Dirac Cone: Bi_2Se_3 and Bi_2Te_3

Soon after the discovery of $Bi_{0.9}Sb_{0.1}$, a new family of stoichiometric crystals, Bi_2Se_3, Bi_2Te_3, and Sb_2Te_3, was identified as three-dimensional topological insulators [10–12]. Among them, Bi_2Se_3 (bismuth selenide) is expected to be the most promising one for applications. It has a large bulk band gap up to 0.3 eV, equivalent to 3,000 K, much higher than room temperatures. Its band inversion happens at the Γ point, leading to a simple band structure of the topological surface states with only single Dirac cone. The high-resolution ARPES measurement shows clearly the surface band dispersion on Bi_2Se_3 as shown in Fig. 7.3, which provides an explicit and unanimous evidence of the surface states of topological insulators. It also reveals a single ring around the $\bar{\Gamma}$ point formed by the pure surface states and the band structure of the Dirac cone. The single Dirac cone of the surface states is now characteristic of topological insulator.

7.1.4   Strained HgTe

Three-dimensional HgTe is a semimetal which is charge neutral when the Fermi level is at the touching point between the light-hole and heavy-hole Γ_8 bands. A unique property of the band structure of HgTe is the inversion of the Γ_6 and Γ_8 band ordering. The effective masses of light and heavy holes have opposite signs (see Fig. 6.5). Appearance of the heavy-hole band between the light-hole and Γ_6 bands makes the material metallic instead of insulating since there is no energy gap in the band structure. Because of the band inversion, three-dimensional HgTe is

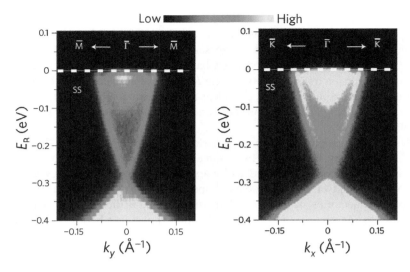

Fig. 7.3 High-resolution ARPES data of surface electronic band dispersion on Bi$_2$Se$_3$(111) measured with an incident photon energy of 22 eV near the Γ point along the $\bar{\Gamma} - \bar{M}$ and $\bar{\Gamma} - \bar{K}$ momentum-space cuts (Reprinted by permission from Macmillan Publisher Ltd: Nature Physics [10], copyright (2009))

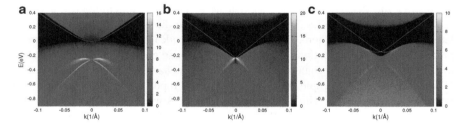

Fig. 7.4 (**a**) The surface local density of states of 3D HgTe without strain, bright line in the direct gap between LH and HH bands indicates the first-type surface states, the bright regimes in valence band indicate the second-type surface states. (**b**) An insulating band gap is opened, and the first- and second-type surface states become connected. (**c**) The surface states when the Γ_6 and the HH band are inverted (Adapted from [14])

also expected to have Dirac-like surface states; however, the surface states always mix with bulk states. Once the system opens an energy gap, it will evolve into a topological insulator. Usually there are two ways to open an energy gap in the band structure. One way is to fabricate a thin film or quantum well. The finite size effect opens a gap, which is the origin of the quantum spin Hall effect in two-dimensional HgTe/CdTe quantum wells. The other way is the strain effect. The strained three-dimensional HgTe is expected to be a topological insulator (see Fig. 7.4). The quantum Hall effect of the surface states in a strained bulk HgTe was observed experimentally [13].

Based on adiabatic continuity of their band structure to HgTe, a series of chalcopyrite semiconductors are predicted to be topological insulators [15].

7.2   Electronic Model for Bi$_2$Se$_3$

Bi$_2$Se$_3$ and Bi$_2$Te$_3$ are three-dimensional topological insulators, which have robust and simple surface states constituting a single Dirac cone at the Γ point [11]. Bi$_2$Se$_3$ and Bi$_2$Te$_3$ share the same rhombohedral crystal structure with the space group D_{3d}^5 (R$\bar{3}$m) with five atoms in one unit cell. We take Bi$_2$Se$_3$ as an example and show its crystal structure in Fig. 7.5, which has layered structure with a triangle lattice within one layer. It has a trigonal axis (threefold rotation symmetry), defined as the z-axis, a binary axis (twofold rotation symmetry), defined as the x-axis, and a bisectrix axis (in the reflection plane), defined as the y-axis. The material consists of five-atom layers arranged along the z-direction, known as quintuple layers. Each quintuple layer consists of five atoms with two equivalent Se atoms (denoted as Se1 and Se1$'$), two equivalent Bi atoms (denoted as Bi1 and Bi1$'$), and a third Se atom (denoted as Se2). The coupling is strong between two atomic layers within one quintuple layer but much weaker, predominantly of the van der Waals type, between

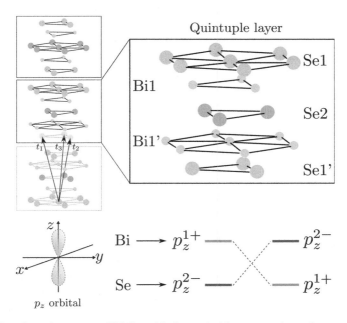

Fig. 7.5 *Top*: Crystal structure of Bi$_2$Se$_3$ with three primitive vectors denoted as $t_{1,2,3}$. The Se1 (Bi1) layer can be related to the Se1$'$ (Bi1$'$) layer by an inversion operation in which the Se2 atoms have the role of inversion centers. *Bottom*: Schematic diagram of the p_z orbitals of electrons, and the band inversion of the p_z^+ orbitals of Bi and the p_z^- orbitals of Se due to the spin-orbit coupling

two quintuple layers. The primitive lattice vectors $t_{1;2;3}$ and rhombohedral unit cells are shown in Fig. 7.5. The Se2 site has the role of an inversion center, and under an inversion operation, Bi1 is changed to Bi1$'$ and Se1 is changed to Se1$'$. The existence of inversion symmetry enables us to construct eigenstates with definite parity for this system.

To get a better understanding of the inversion of the band structure and the parity exchange in Fig. 7.5, we start from the atomic energy levels and consider the effect of crystal-field splitting and spin-orbit coupling on the energy eigenvalues at the Γ point. The states near the Fermi surface mainly come from p orbitals of Bi $(6s^2 6p^3)$ and Se $(4s^2 4p^4)$. The crystal field removes the degeneracy of the p orbitals, and only the p_z orbitals become relevant in the present problem. Furthermore, due to the inversion symmetry of the crystal lattice, the p_z orbitals of electrons from Bi and Se atoms near the Fermi surface have opposite parity. The band gap between these two orbitals is controlled by the spin-orbit coupling: increasing the spin-orbit coupling may cause a band inversion as analyzed in [11].

The three-dimensional Dirac equation can be applied to describe a large family of three-dimensional topological insulators. $Bi_2 Te_3$, $Bi_2 Se_3$, and $Sb_2 Te_3$ have been confirmed to be topological insulators with a single Dirac cone of surface states. For example, in $Bi_2 Te_3$, the electrons near the Fermi surfaces, mainly come from the p orbitals of Bi and Te atoms. According to the point group symmetry of the crystal lattice, p_z orbital splits from $p_{x,y}$ orbitals. Near the Fermi surface the energy levels turn out to be the p_z orbital, $|P1_z^+, \uparrow\rangle$, $|P1_z^+, \downarrow\rangle$, $|P2_z^-, \uparrow\rangle$, and $|P2_z^-, \downarrow\rangle$, where $\pm$ stand for the parity of the corresponding states and $\uparrow, \downarrow$ for the electron spin. Four low-lying states at the Γ point can be used a basis to construct the low-energy effective Hamiltonian [11]. In the basis of $(|P1_z^+, \uparrow\rangle, |P1_z^+, \downarrow\rangle, |P2_z^-, \uparrow\rangle, |P2_z^-, \downarrow\rangle)$, we keep the terms up to the quadratic order in p and obtain

$$H = \epsilon(p) + \sum_{i=x,y,z} v_i p_i \alpha_i + \left(M - \sum_{i=x,y,z} B_i p_i^2 \right) \beta \qquad (7.1)$$

with $v_x = v_y = v_\parallel$ and $v_z = v_\perp$ and $B_x = B_y = B_\parallel$ and $B_z = B_\perp$. The first term $\epsilon(p) = C - D_\parallel(p_x^2 + p_y^2) - D_\perp p_z^2$ which breaks the particle-hole symmetry of the system. The linear term in p_i is determined by the change of parity of the different basis. Anisotropy of the crystal reveals that $B_\parallel \neq B_\perp$ and $v_\parallel \neq v_\perp$. It will modify the effective velocity of the surface states.

This model can be understood as a result of $k \cdot p$ theory. Under the time reversal, $\alpha_i \to -\alpha_i$ and $\beta \to \beta$. Suppose the system is time reversal invariant. Expand the Hamiltonian near $p = 0$. The zero-order term is $M\beta$ where M represents the energy gap at the point, the first-order term is $\sum_{i=x,y,z} v_i p_i \alpha_i$ since $p_i \to -p_i$, and the second-order term is $\sum_{i=x,y,z} B_i p_i^2 \beta$ since $p_i^2 \to p_i^2$ under the time reversal. $\epsilon(p)$ is the dispersion independent of inter-band coupling.

7.3 Effective Model for Surface States

The effective Hamiltonian for the surface states can be derived from the electronic model for the bulk. Consider an x-y plane at $z = 0$. p_x and p_y are good quantum numbers, and p_z is replaced by $-i\hbar\partial_z$ in Eq. (7.1) To derive an effective Hamiltonian for the surface states, we first find the solution of the surface states at $p_x = p_y = 0$ in Eq. (7.1),

$$H(z)\,|\Psi\rangle = E\,|\Psi\rangle \tag{7.2}$$

where

$$H(z) = C + D_\perp \hbar^2 \partial_z^2 - iv_\perp \hbar \partial_z \alpha_z + (M + B_\perp \hbar^2 \partial_z^2)\beta. \tag{7.3}$$

We have derived the effective model for surface states for the modified Dirac equation in Chap. 2. Here the presence of $\epsilon(p)$ makes the problem a little bit more complicated to find the solution at $p_x = p_y = 0$. The term $\epsilon(p)$ breaks the particle-hole symmetry between the conduction band and valence band. If $D_\perp^2 > B_\perp^2$, the band gap closes and the system is no longer an insulator. To have a surface state solution, we focus on the case of $D_\perp^2 < B_\perp^2$. In this matrix equation, the first and third rows are decoupled from the second and fourth rows. For this reason, we can adopt two trial wave functions:

$$|\Psi_1\rangle = \begin{pmatrix} a_1 \\ 0 \\ b_1 \\ 0 \end{pmatrix} e^{\lambda z}, |\Psi_2\rangle = \begin{pmatrix} 0 \\ a_2 \\ 0 \\ b_2 \end{pmatrix} e^{\lambda z}, \tag{7.4}$$

respectively. The equation is reduced into two independent sets of equations:

$$\begin{pmatrix} M + B_+\lambda^2 & -iv_\perp\lambda \\ -iv_\perp\lambda & -M + B_-\lambda^2 \end{pmatrix} \begin{pmatrix} a_1 \\ b_1 \end{pmatrix} = E \begin{pmatrix} a_1 \\ b_1 \end{pmatrix} \tag{7.5}$$

and

$$\begin{pmatrix} M + B_+\lambda^2 & +iv_\perp\lambda \\ +iv_\perp\lambda & -M + B_-\lambda^2 \end{pmatrix} \begin{pmatrix} a_2 \\ b_2 \end{pmatrix} = E \begin{pmatrix} a_2 \\ b_2 \end{pmatrix} \tag{7.6}$$

where $B_\pm = B_\perp \pm D_\perp$. These two equations are equivalent to those for the edge states in the quantum spin Hall effect at $k_x = 0$. We first focus on the solution of a_1 and b_1. To have a nontrivial solution, the characteristic equation is

$$\det \begin{pmatrix} M + B_+\lambda^2 - E & -iv_\perp\lambda \\ -iv_\perp\lambda & -M + B_-\lambda^2 - E \end{pmatrix} = 0. \tag{7.7}$$

From this equation, we find four roots for λ: $\pm\lambda_1$ and $\pm\lambda_2$. We adopt the Dirichlet boundary conditions, which require that the wave function for the surface states must vanish at $z = 0$ and $z \to -\infty$. For $MB_\perp > 0$, we obtain

$$
|\Psi_1\rangle = \begin{pmatrix} a_1 \\ 0 \\ b_1 \\ 0 \end{pmatrix} \left(e^{\lambda_1 z} - e^{\lambda_2 z} \right) \tag{7.8}
$$

with

$$
\lambda_1 = \frac{1}{2} \sqrt{\frac{v_\perp^2}{B_+ B_-}} + \sqrt{\frac{1}{4} \frac{v_\perp^2}{B_+ B_-} - \frac{M}{B_\perp}}, \tag{7.9a}
$$

$$
\lambda_2 = \frac{1}{2} \sqrt{\frac{v_\perp^2}{B_+ B_-}} - \sqrt{\frac{1}{4} \frac{v_\perp^2}{B_+ B_-} - \frac{M}{B_\perp}}, \tag{7.9b}
$$

which requires that

$$
\frac{a_1}{b_1} = \frac{i A \lambda_1}{B_+(\lambda_1^2 + \frac{M}{B_\perp})} = \frac{i A \lambda_2}{B_+(\lambda_2^2 + \frac{M}{B_\perp})}. \tag{7.10}
$$

The normalization of the wave function requires

$$
|a_1|^2 + |b_1|^2 = \left(\frac{\lambda_1 + \lambda_2}{2\lambda_1 \lambda_2} - \frac{2}{\lambda_1 + \lambda_2} \right)^{-1}. \tag{7.11}
$$

Similarly, we find the solution to $|\Psi_2\rangle$ by setting $a_2 = -a_1$ and $b_2 = b_1$:

$$
|\Psi_2\rangle = \begin{pmatrix} 0 \\ -a_1 \\ 0 \\ b_1 \end{pmatrix} \left(e^{\lambda_1 z} - e^{\lambda_2 z} \right). \tag{7.12}
$$

The energy eigenvalue for both states is $E = -D_\perp M / B_\perp$.

To find the solution of $p_x, p_y \neq 0$, we may use the projection and perturbation method by utilizing the two solutions at $p_x = p_y = 0$ as the basis to project the Hamiltonian. On the new basis, the effective Hamiltonian is projected out:

$$
H_{\text{eff}} = \begin{pmatrix} \langle \Psi_1 | H | \Psi_1 \rangle & \langle \Psi_1 | H | \Psi_2 \rangle \\ \langle \Psi_2 | H | \Psi_1 \rangle & \langle \Psi_2 | H | \Psi_2 \rangle \end{pmatrix}. \tag{7.13}
$$

In this way, we obtain an effective Hamiltonian in the x-y plane [16]

$$
H_{\text{eff}} = \epsilon_0(p) + v_{\text{eff}}(p \times \sigma)_z \tag{7.14}
$$

where $v_{\text{eff}} = \text{sgn}(B_\perp)\sqrt{1 - D_\perp^2/B_\perp^2}\,v_\parallel$. We note that the inclusion of $\epsilon(p)$ will revise the effective velocity of the surface states. A quadratic term appears up to p^2, $\epsilon_0(p) = E_0 - D_\parallel(p_x^2 + p_y^2)$. Note that the effective Hamiltonian is only valid for a small p.

A key feature of this effective model is the lock-in relation between the momentum and spin. In the polar coordinate, we set $p = \sqrt{p_x^2 + p_y^2}$ and $\phi_p = \arctan p_y/p_x$. The dispersions for the surface states are $E_\pm(p) = \epsilon_0(k) \pm v_{\text{eff}}p$, and the corresponding energy states are

$$|\Psi_\pm\rangle = \frac{1}{\sqrt{2}}\begin{pmatrix} \pm e^{-i\phi_p} \\ i \end{pmatrix}. \tag{7.15}$$

The Berry phase, which is acquired by a state upon being transported around a loop in the k space, can be evaluated exactly:

$$\gamma_\pm = \oint d\phi_p \langle \Psi_\pm | i\frac{\partial}{\partial \phi_p} | \Psi_\pm \rangle = \pi. \tag{7.16}$$

The Berry phase will play an essential role in transport properties of the surface states, such as weak antilocalization. An ideal Dirac fermion gas is a super-metal, in which none of the states can be localized by disorders or impurities.

Hexagonal Warping Effect [17]: Bi_2Te_3 has a rhombohedral crystal structure with space group $R3\bar{m}$. In the presence of a [111] surface, the crystal symmetry is reduced to C_{3v}, which consists of a threefold rotation C_3 around the trigonal z-axis and a mirror operation M: $x \to -x$ where x is in $\Gamma - K$ direction. Under the operations of C_3 and M, the momentum and spin transform as follows:

$$C_3 : p_\pm \to e^{\pm i2\pi/3}p_\pm, \sigma_\pm \to e^{\pm i2\pi/3}\sigma_\pm, \sigma_z \to \sigma_z, \tag{7.17a}$$

$$M : p_+ \to -p_-, \sigma_x \to \sigma_x, \sigma_{y,z} \to -\sigma_{y,z}. \tag{7.17b}$$

In addition, time reversal symmetry gives the constraint

$$H(\mathbf{p}) = \Theta H(-\mathbf{p})\Theta^{-1}. \tag{7.18}$$

Keeping the higher-order term up to p^3, the effective Hamiltonian for the surface states has the form

$$H_{\text{eff}} = \epsilon_0(p) + v_{\text{eff}}(p_x\sigma_y - p_y\sigma_x) + \frac{\lambda}{2}(p_+^3 + p_-^3)\sigma_z. \tag{7.19}$$

where $\epsilon_0 = p^2/2m^*$. The cubic term does not break time reversal symmetry.

7.4 Physical Properties of Topological Insulators

7.4.1 Absence of Backscattering

The absence of backscattering in the topological surface states can be demonstrated as following: a pair of the Kramers' states $|k, \uparrow\rangle$ and $|-k, \downarrow\rangle$ are related by the time reversal transformation, $|-k, \downarrow\rangle = \Theta |k, \uparrow\rangle$. Since the operator Θ is anti-unitary, it is straightforward that

$$\langle -k, \downarrow | U | k, \uparrow \rangle = - \langle -k, \downarrow | \Theta U \Theta | k, \uparrow \rangle$$
$$= - \langle k, \uparrow | U | -k, \downarrow \rangle^* = - \langle -k, \downarrow | U | k, \uparrow \rangle, \qquad (7.20)$$

where U is a time reversal invariant operator. Thus for a potential of a nonmagnetic impurity V, $\langle -k, \downarrow | V | k, \uparrow \rangle = 0$.

The absence of backscattering of the surface states was studied in the alloy $Bi_{1-x}Sb_x$ [18] and in the single crystal Bi_2Te_3 [19]. Bi_2Te_3 has only a single Dirac cone and therefore clearer picture. The constant energy contour at the Fermi energy of the conduction band of Bi_2Te_3 is shown in Fig. 7.6. Due to strong warping effect in Eq. (7.19), the constant energy contour of the surface band of Bi_2Te_3 is not a perfect ring but looks like a hexagram. In the scanning tunneling microscopy (STM) measurement on the surface of Bi_2Te_3, nonmagnetic Ag atom trimers are deposited, which can scatter the surface states. The electron wave functions before and after scattering will interfere with each other and form a standing wave pattern. The fast Fourier transformation from the real-space standing wave pattern to the momentum space can reveal the momentum difference before and after the scattering. On the hexagram, the density of states is not uniform. At some momenta, the density of states is relatively large, as depicted by the darker area on the hexagram in Fig. 7.6. The scattering between these high-density momenta is more obvious than others. The momentum difference between two momenta with totally opposite momenta, that is, backscattering $\mathbf{q} = \mathbf{k}_f - \mathbf{k}_i = 2\bar{K}$. If the backscattering is present, there will

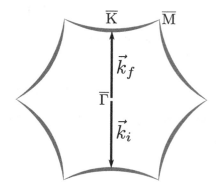

Fig. 7.6 The constant energy contour of H_{eff} in Eq. (7.19) The dominant scattering wave vectors connect two points in $\bar{\Gamma}$-$\bar{K}$ directions on constant energy contour. $\mathbf{k}_i$ and $\mathbf{k}_f$ denote the wave vectors of incident and scattered states

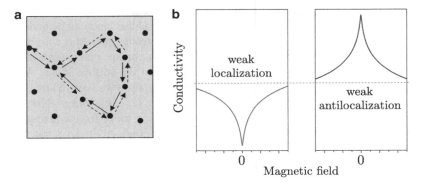

Fig. 7.7 (**a**) The backscattering between two time-reversed scattering loops. (**b**) The exhibition of weak localization and weak antilocalization in magnetoconductivity. The *horizontal dashed line* marks the classical conductivity

be a high-value signal along **q** direction in the fast Fourier transformation spectra, more specifically, along the $\bar{K}$ direction of the surface Brillouin zone. However, there are apparent gaps along the $\bar{K}$ direction in the STM measurement [19]. This provides a direct evidence of the absence of backscattering of Dirac fermions on the surface of topological insulator.

7.4.2   Weak Antilocalization

The weak antilocalization is a characteristic feature in transport experiments that demonstrates the presence of the Dirac fermions in topological insulators. It appears as the low-field negative magnetoconductivity, that is, negative conductivity change as a function of applied magnetic field [20–24]. A series of experimental measurements were reported. So far, all the reported samples in the transport experiments have low mobility and short mean free path, so the diffusion dominates the electronic transport. Like many semiconductors, the phase coherence length in topological insulators can be as long as several hundred nanometers to more than 1 μm at low temperatures (below the liquid helium temperature). When the sample size is comparable with the phase coherence length, the quantum interference becomes an important correction to the diffusion transport. In materials without or with ignorable spin-orbit coupling, the constructive quantum interference will enhance the backscattering between two time-reversed scattering loops (Fig. 7.7) and suppress the resistivity. This suppression of resistivity by the quantum interference leads to the weak localization. It can be destroyed by applying a magnetic field that breaks the constructive quantum interference. On the other hand, it has been long time since people had realized that strong spin-orbit scattering in some solids could also make the quantum interference change from constructive to destructive. As a result, the conductivity is enhanced and weak antilocalization happens.

Besides the spin-orbit scattering, the energy band structure with spin-orbit coupling can also lead to the weak antilocalization, and this case can be understood in terms of the Berry phase argument. Due to the strong spin-orbit coupling, the surface states of topological insulator have a two-component spinor wave function, which describes a momentum-spin lock-in relation of the surface states. After an electron circled around the Dirac point, its spin orientation was rotated by 2π, and the wave function accumulates only a π Berry phase [25, 26]. The π Berry phase changes the interference of the time-reversed scattering loops from constructive to destructive. The destructive interference will give the conductivity an enhancement, which can be destroyed by applying a magnetic field, leading to the negative magnetoconductivity with the cusp.

7.4.3   Shubnikov-de Haas Oscillation

All of early in-plane transport measurements reveal the dominance of the three-dimensional bulk conductivity [20, 21, 27]. One way to determine the dimension of the carriers and to distinguish the two-dimensional surface states from the three-dimensional bulk states is the Shubnikov-de Haas oscillation. In the presence of a strong perpendicular magnetic field, an electron gas splits into discrete Landau levels; the separation between the Landau levels increases with the increasing magnetic field. As the magnetic field increases, the Landau levels cut through the Fermi surface one by one. When the Fermi level is (not) aligned with a Landau level, the resistivity drops (increases). As a result, the in-plane measurement will measure an oscillating resistivity, known as the Shubnikov-de Haas oscillation. Because Shubnikov-de Haas oscillation only responds to a perpendicular magnetic field, a two-dimensional electron gas has no Shubnikov-de Haas oscillation for in-plane magnetic fields, while a three-dimensional electron gas can have Shubnikov-de Haas oscillation for magnetic field applied along any directions. This makes the angle dependence of Shubnikov-de Haas oscillation a convenient tool to identify the dimension of carriers. Shubnikov-de Haas oscillation revealed the coexistence of three-dimensional bulk carriers with the two-dimensional surface states in the transport for $Bi_{1-x}Sb_x$ [27, 28] and Bi_2Se_3 [29–31]. The Shubnikov-de Haas oscillation measured in a Bi_2Se_3 crystal shows that the bulk states dominate the transport, because it can be measured for arbitrary magnetic field direction. Shubnikov-de Haas oscillation also revealed the Berry phase information. The oscillating longitudinal resistivity ρ_{xx} can be formulated as

$$\rho_{xx} \sim \cos\left[2\pi\left(\frac{F}{B} - \gamma\right)\right], \tag{7.21}$$

where F is the oscillation frequency and γ is the phase of the oscillation. The Berry phase can be found as [32] $2\pi(\gamma - \frac{1}{2})$. One has zero Berry phase for $\gamma = \frac{1}{2}$ and π

Berry phase for $\gamma = 0$. The Berry phase is about 0.4, giving another signature that the bulk states dominate the transport of as-grown topological insulator Bi_2Se_3.

7.5   Surface Quantum Hall Effect

When the surface states are subjected to a Zeeman field, the massless Dirac fermions gain a mass and open an energy gap,

$$H = v(p \times \sigma)_z + \Delta\sigma_z = d \cdot \sigma \qquad (7.22)$$

with $d_x = -vp_y$, $d_y = vp_x$, and $d_z = \Delta$. From the Kubo formula, the Hall conductance can be expressed as

$$\sigma_{xy} = \frac{e^2}{2\hbar} \int \frac{dk_x dk_y}{(2\pi)^2} \frac{(f_{k,+} - f_{k,-})\, \mathbf{d}(k) \cdot \partial_{k_x}\mathbf{d}(k) \times \partial_{k_y}\mathbf{d}(k)}{|\mathbf{d}(k)|^3} \qquad (7.23)$$

where $f_{k,\pm} = \{1 + \exp[(\pm |\mathbf{d}(k)| - \mu)/k_B T]\}^{-1}$ (for details see Sect. A.2). When the Fermi energy level is located in the gap, that is, $\mu = 0$, the Hall conductance is half quantized at zero temperature:

$$\sigma_{xy} = -\frac{\text{sgn}(\Delta)}{2}\frac{e^2}{h}. \qquad (7.24)$$

It is noted that the Hall conductance is usually related to the Chern number, which is always an integer if the Brillouin zone is finite as we prove in Sect. A.1. However, here the integral range is infinite, which makes it possible that the conductance is not an integer.

This is regarded as one of the key features of the surface states in topological insulator. It has a lot of applications in the field of topological insulator. For example, it plays a decisive role in the development of topological field theory [33].

Although we have a half quantized Hall conductance from the Kubo formula, it is still not clear whether or not the half quantization of the Hall conductance can be directly observed in transport measurement. In the integer quantum Hall system, the current-carrying chiral edge states are responsible for the quantized conductance measured in transport experiment [34, 35]. It is not immediately clear whether or not the similar chiral edge state will form on the closed surface of a topological insulator and how the quantized nature of the edge states can be reconciled with the prediction of the half quantization of the Hall conductance [33, 36, 37]. To get a definite answer to these questions, we investigate the multiterminal transport properties of a three-dimensional topological insulator in the presence of a uniform spin-splitting Zeeman field.

To illustrate the basic physics, we consider a three-dimensional topological insulator of the cubic shape. A Zeeman field is applied along the z-direction, as

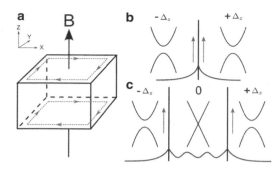

Fig. 7.8 (a) Schematic of a three-dimensional topological insulator in a weak Zeeman field and the formation of the chiral current on the top and bottom surface boundary. (b) A chiral edge state around the domain wall between the two-dimensional Dirac fermions with the positive and negative masses. (c) An edge mode is splitted into two halves separating by a metallic area (Adapted from [42])

shown in Fig. 7.8a. Because the bulk of the system is insulating, it is effectively a closed two-dimensional surface with six facets. The effective Hamiltonian of the Dirac fermions for the surface states can be written as [16, 38]

$$H_{\text{eff}}(\mathbf{k}) = v\,(\mathbf{k} \times \sigma) \cdot \mathbf{n} - g_\parallel \mu_B h_\parallel \sigma_\parallel - g_\perp \mu_B \mathbf{h}_\perp \cdot \sigma_\perp, \qquad (7.25)$$

where $\mathbf{n}$ denotes the normal vector of the surface, $\sigma \equiv \{\sigma_x, \sigma_y, \sigma_z\}$ are the Pauli matrices, $h_\parallel$ ($\sigma_\parallel$) and $h_\perp$ ($\sigma_\perp$) are the Zeeman field (Pauli matrix) components parallel and perpendicular to the normal vector, respectively, and $g_\parallel$ and $g_\perp$ are the corresponding spin g-factors. Note that the surface states have anisotropic spin g-factors due to the strong spin-orbit coupling of the bulk band: $g_\parallel$ is the same as that of the bulk material, and $g_\perp$ is renormalized by bulk band parameters and is usually strongly suppressed [16, 38]. Different facets of the surface have different effective Hamiltonian respective to different normal vectors $\mathbf{n}$. For the top and the bottom facets, the effective Hamiltonian can be written as

$$H_{\text{eff}} = \pm v(k_x \sigma_y - k_y \sigma_x) + \Delta_z \sigma_z, \qquad (7.26)$$

where $+$ and $-$ are for the top and bottom surfaces, respectively, and $\Delta_z \equiv -g_\parallel \mu_B h$. The spectrum will open a gap on these facets, and the Dirac fermions gain a mass $\pm \Delta_z$. On the other hand, the effective Hamiltonian for the side facets can be written as

$$H_{\text{eff}} = v[(k_x + \Delta_h)\sigma_z - k_z \sigma_x], \qquad (7.27)$$

where $\Delta_h \equiv g_\perp \mu_B h$. In this case, the Zeeman field simply shifts the Dirac point from ($k_x = 0, k_y = 0$) to ($-\Delta_h, 0$). When the Fermi level is located in the gap of the top and bottom surface, the system becomes effectively two insulating domains separated by a conducting belt with massless Dirac fermions.

Fig. 7.9 Schematic illustration of the three-dimensional (3D) device with two-dimensional (2D) semi-infinite metallic leads

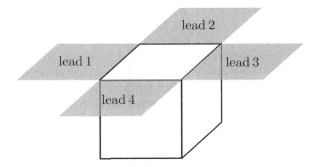

A chiral edge state will form and is concentrated around the boundaries between the insulating domains and the metallic belt as illustrated in Fig. (7.8)b, c. Effectively, the chiral edge state is split into two halves, each of which is circulating around the boundary of one of the domains and carrying one half of the conductance quantum e^2/h. Such a picture reconciles the apparent conflict between the half quantization and the index theorem. After establishing the existence of the chiral surface-edge states, we can calculate the Hall conductance numerically using the Landauer-Büttiker formalism [39–41]. The setup of the device is illustrated in Fig. 7.9; four identical two-dimensional metallic leads ($\mu = 1, 2, 3, 4$) are attached to the top square surface of a semi-infinite three-dimensional topological insulator, acting as the measurement electrodes. The Zeeman field is normal to the top surface, and the Fermi level is located in the gap. The multiterminal conductance can be deduced from the transmission coefficient T_{pq} from the terminal p to terminal q,

$$T_{pq} = \text{Tr}[\Gamma_p G^r \Gamma_q G^a] \qquad (7.28)$$

where Γ_p is determined by the self-energy at the terminal p [40]. The advanced and retarded Green's functions are given by

$$G^{R,A}(E) = \frac{1}{E - H_c - \sum_{p=1}^{4} \Pi_p^{R,A}}, \qquad (7.29)$$

where E is the electron energy and H_c is the model Hamiltonian for three-dimensional system. The retarded and advanced self-energy terms are introduced for the semi-infinite lead p [41].

In this way, the transmission coefficients as a function of the thickness of the sample can be calculated numerically. When the sample is thick enough, it was found that the transmission coefficients between the two neighboring terminals p and q have the relation [42]

$$T_{pq} - T_{qp} = \frac{1}{2}. \qquad (7.30)$$

A straightforward way to measure the "half quantized" Hall conductance in the four-terminal setup is to apply a voltage between the terminals 1 and 3 (V_{13}) and measure the current between the terminals 2 and 4 (I_{24}). It is easy to show that the cross conductance $\sigma_{24,13} \equiv I_{24}/V_{13} = (e^2/2h)(T_{12} - T_{21})$, yielding $e^2/4h$ for the half quantization. The measurement using the usual six-terminal Hall bar configuration could be more tricky due to the presence of the metallic side facets, which give rise to the finite longitudinal conductance σ_L. In the limit of thick sample with $\sigma_L \gg e^2/h$, the Hall conductance σ_H approaches $(4e^2/h)(T_{pq} - T_{qp})$ (if we assume $T_{pq} - T_{qp}$ is the same between all neighboring leads), which yields $2e^2/h$ for the half quantization. It can be compared with the case of the quantum Hall effect where σ_L vanishes when σ_H shows quantization [42].

7.6 Surface States in a Strong Magnetic Field

We come to study the surface states subjected to a uniform magnetic field. We first consider a geometry of strip with width L_y and thickness H, which are much larger than the magnetic length l_B and the spatial distribution ξ of the surface states. Suppose the magnetic field B (alone the z-axis) is perpendicular to the slab. We focus on the top plane. The periodic boundary condition is adopted along the x-axis and the open boundary condition along the y-axis. In this way, the wave number k_x is still a good quantum number, and k_y is substituted by $-i\partial_y$. We take the Landau gauge for the vector potential, $A_x = -By$ and $A_y = 0$. In this way, the effective model Eq. (7.22) in a B field can be expressed in

$$H_{\text{eff}} = v_F \left[(\hbar k_x - eBy)\sigma_y + i\hbar\partial_y\sigma_x \right] + \Delta\sigma_z. \tag{7.31}$$

To solve the problem, it is convenient to define

$$a(y_0) = i\frac{l_B}{\sqrt{2}}[\partial_y + l_B^{-2}(y - y_0)] \tag{7.32}$$

where the magnetic length $l_B = \sqrt{\hbar/eB}$ and $y_0 = l_B^2 k_x$ assuming $eB > 0$. The defined operators a and $a^\dagger$ satisfy the commutation relation,

$$[a(y_0), a^\dagger(y_0)] = 1. \tag{7.33}$$

For simplicity, we introduce a dimensionless parameters $m_0 = \frac{\Delta}{\sqrt{2}\hbar v_F/l_B}$. In this way, we obtain a dimensionless Schrödinger equation:

$$\begin{pmatrix} m_0 & a \\ a^\dagger & -m_0 \end{pmatrix} \begin{pmatrix} \varphi_1 \\ \varphi_2 \end{pmatrix} = \frac{E}{v_F\sqrt{2e\hbar B}} \begin{pmatrix} \varphi_1 \\ \varphi_2 \end{pmatrix}. \tag{7.34}$$

The allowed values for y_0 are separated by $\delta y_0 = 2\pi l_B^2 / L_x$ for a periodic boundary condition with length L_x and are limited within $0 < y_0 < L_y$. The solution is a function of the good quantum number k_x or $y_0 = l_B^2 k_x$. When y_0 is far away from two edges of $y = 0$ and $y = L$, the two components φ_1 and φ_2 will vanish at the two boundaries. Let $|0\rangle$ the lowest energy state for simple harmonic oscillator such that $a(y_0) |0\rangle = 0$. $|n\rangle = \frac{1}{(n!)^{1/2}} (a^\dagger(y_0))^n |0\rangle$ is the eigenstates of $N(y_0) = a^\dagger(y_0) a(y_0)$ with eigenvalue n (an integer). In this case, the energy eigenstates in Eq. (7.34) are

$$|n, \alpha\rangle = \begin{pmatrix} \sin\theta_{n,\alpha} |n-1\rangle \\ \cos\theta_{n,\alpha} |n\rangle \end{pmatrix} \tag{7.35}$$

where $\tan\theta_{n,\alpha} = \frac{\sqrt{n}}{\alpha\sqrt{n+m_0^2}-m_0}$ and $\alpha = \pm 1$ [43]. The Landau energy is given by

$$E_{n,\alpha} = \alpha v_F \sqrt{2e\hbar B \left(n + m_0^2\right)}, \tag{7.36}$$

which are highly degenerate for different values of y_0. The number of the allowed values of y_0, $N_L = L_y/\delta y_0 = 2\pi L_x L_y / l_B^2$, which is called the degeneracy of the Landau levels.

It should be emphasized that the zero mode $E_0 = -v_F \sqrt{2e\hbar B} |m_0|$ for $n = 0$ and the eigenstate is fully saturated, $|0, 0\rangle = \begin{pmatrix} 0 \\ |0\rangle \end{pmatrix}$. The energy expressions yield an energy gap $\Delta E = |E_{n=\pm 1}| - E_0$ between the zero mode and the states of $n = \pm 1$. For $m_0 = 0$, the energy gap is about $\Delta E \approx 800\,\text{K}$ for Bi_2Se_3 at $B = 10\,\text{T}$, which makes it possible that the quantum Hall effect can be measured even at room temperature just as in single layer graphene [44].

Unlike the conventional two-dimensional electron gas where the Landau levels are evenly spaced $E_n = \hbar\omega_c(n + \frac{1}{2})$ and the lowest Landau level of the conventional two-dimensional electron gas has a nonzero energy $\hbar\omega_c/2$ (ω_c is the cyclotron frequency), the Landau-level energies of massless Dirac fermions have a square-root dependence on magnetic field B and the level index n, given by

$$E_n = \text{sgn}(n) v_F \sqrt{2eB\hbar |n|}, \tag{7.37}$$

where the level index $n = 0, \pm 1, \pm 2, \ldots$. Moreover, Dirac fermions can have zero Landau-level index $n = 0$, even negative level indices $n < 0$. This square-root dependence has been observed in the measurement of scanning tunneling spectroscopy [45, 46]. Despite the observation of the Landau levels in the STM measurement, an in-plane measurement with a Hall-bar setup still poses an experimental challenge, so the quantum Hall conductance has not yet been observed for the surface states of topological insulator.

7.7   Topological Insulator Thin Film

Thin film of three-dimensional topological insulator may provide an alternative way to realize the quantum spin Hall effect. It is an example to reduce a three-dimensional topological insulator to two-dimensional topological insulator. The surface states have spatial distribution, which can be characterized by a length scale ξ_s. When this length scale is comparable with the thickness of the thin film, the wave functions of the two surface states from top and bottom surfaces will overlap in space. Consequently, the two surface states open gaps. Thus the surface states of the thin film can be described by a two-dimensional massive Dirac model [16, 38].

7.7.1   Effective Model for Thin Film

Consider an extra-thin film in the x-y plane such that k_x and k_y are good quantum numbers, and the thickness of the thin film along z-direction is denoted as L. To establish an effective model for an ultrathin film, we still start with the electronic model in Eq. (7.1) and follow the approach to derive the effective model for the surface states where only one surface is considered in Sect. 7.3. The boundary condition in the present problem is different since two surfaces should be considered simultaneously. If the thin film is so thick that the surface states at the top and bottom layers are well separated, that is, $L \gg \lambda_1^{-1}, \lambda_2^{-1}$ the characteristic scales of the surface states defined in Eq. (7.9), the thin film consists of two independent massless Dirac cones when the Fermi level is located in the bulk gap. However, if the thickness L is comparable with λ_1^{-1} and λ_2^{-1}, the two surface states at the top and bottom layers will be coupled together and open an energy gap at the Dirac point. Thus the massless Dirac electrons gain a mass and evolve into massive Dirac electrons.

At $k_x = k_y = 0$, we have four roots for λ in Eq. (7.7): $\pm\lambda_1$ and $\pm\lambda_2$ as functions of energy E. Thus the final solution for the wave function should be a linear superposition of these solutions, for example,

$$|\Psi_1\rangle = \sum_{i=1}^{4} c_i \begin{pmatrix} a_i \\ 0 \\ b_i \\ 0 \end{pmatrix} e^{\lambda_i z}. \tag{7.38}$$

We take the Dirichlet boundary condition for the wave functions at $z = \pm L/2$, that is, $\Psi(z = \pm\frac{L}{2}) = 0$. Then we can obtain a set of transcendental equations to determine the values of E, λ_1, and λ_2 as function of thickness L:

$$\frac{\alpha_1^2 \lambda_2^2 + \alpha_2^2 \lambda_1^2}{\alpha_1 \alpha_2 \lambda_1 \lambda_2} = \frac{\tanh\frac{\lambda_1 L}{2}}{\tanh\frac{\lambda_2 L}{2}} + \frac{\tanh\frac{\lambda_2 L}{2}}{\tanh\frac{\lambda_1 L}{2}}, \tag{7.39}$$

note that where $\alpha_{1,2} = E - C - M - (D_1 + B_1)\lambda_{1,2}^2$. In Eq. (7.39), λ_α define the behavior of the wave functions along z-axis and are functions of the energy E

$$\lambda_\alpha(E) = \sqrt{\frac{-F + (-1)^{\alpha-1}\sqrt{R}}{2(D_\perp^2 - B_\perp^2)}}, \tag{7.40}$$

where for convenience we have defined $F = A_\perp^2 + 2D_\perp(E - C) - 2B_\perp M$ and $R = F^2 - 4(D_\perp^2 - B_\perp^2)[(E - C)^2 - M^2]$. The equations in (7.39) and (7.40) can be solved numerically, and give two energies at the Γ point, that is, E_+ and E_-, which define an energy gap

$$\Delta \equiv E_+ - E_-. \tag{7.41}$$

We can find two solutions of $|\Psi_1\rangle$: φ^1 for E_+ and χ^1 for E_- and other two solutions for $|\Psi_2\rangle$: φ^2 for E_+ and χ^2 for E_-. For details, readers may refer to the references [16, 38]. By using these four solutions as basis states and rearranging their sequence following (note that each basis state is a four-component vector) $(\varphi^1, \chi^2, \chi^1, \varphi^2)$, we can map the original Hamiltonian to the Hilbert space spanned by these four states and reach a new low-energy effective Hamiltonian for the ultrathin film,

$$H_{\text{eff}} = \begin{bmatrix} h_+(k) & 0 \\ 0 & h_-(k) \end{bmatrix} \tag{7.42}$$

in which

$$h_{\tau_z}(k) = E_0 - Dk^2 - \hbar v_F(k_x\sigma_y - k_y\sigma_x) + \tau_z\left(\frac{\Delta}{2} - Bk^2\right)\sigma_z. \tag{7.43}$$

Note that here the basis states of Pauli matrices stand for spin-up and spin-down states of real spin. In Eq. (7.43), we have introduced a hyperbola index $\tau_z = \pm 1$ (or $\pm$). Unlike the momentum correspondence in graphene, it is the σ_z to $-\sigma_z$ correspondence in the present case. Therefore, the dispersions of $h_\pm$ are actually doubly degenerate, which is secured by time reversal symmetry. Here, $\tau_z = \pm$ are used to distinguish the two degenerate hyperbolas, $h_+(k)$ and $h_-(k)$ describe two sets of Dirac fermions, and each show a pair of conduction and valence bands with the dispersions:

$$\varepsilon_\pm(\mathbf{k}) = E_0 - Dk^2 \pm \sqrt{\left(\frac{\Delta}{2} - Bk^2\right)^2 + (\hbar v_F)^2 k^2}, \tag{7.44}$$

where $+$ and $-$ correspond to the conduction and valence bands, respectively. The eigenstates for $\varepsilon_\pm$ are

$$u_\pm(\mathbf{k}) = \frac{1}{\|u_\pm\|}\begin{bmatrix} \left(\frac{\Delta}{2} - Bk^2\right)\tau_z \pm \sqrt{\left(\frac{\Delta}{2} - Bk^2\right)^2 + (\hbar v_F)^2 k^2} \\ -i\hbar v_F k_+ \end{bmatrix} \tag{7.45}$$

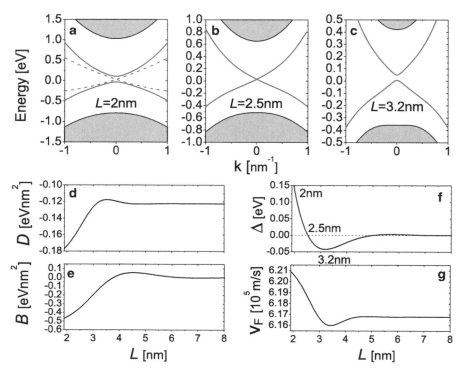

Fig. 7.10 (a)–(c) Twofold degenerate ($\tau_z = \pm 1$) energy spectra of surface states for thicknesses $L = 20, 25, 32$ Å (*solid lines*), and $L \to +\infty$ (*dash lines*). The *gray* area corresponds to the bulk states. The parameters are $M = 0.28$ eV, $A_1 = 2.2$ eVÅ, $A_2 = 4.1$ eVÅ, $B_1 = 10$ eVÅ^2, $B_2 = 56.6$ eVÅ^2, $C = -0.0068$ eV, $D_1 = 1.3$ eVÅ^2, and $D_2 = 19.6$ eVÅ^2. (**d**)–(**g**) The calculated parameters for the new effective model H_{eff} as a function of thickness L (Adapted from [38])

with

$$\|u_{\pm}\| = \sqrt{\left[\left(\frac{\Delta}{2} - Bk^2\right)\tau_z \pm \sqrt{\left(\frac{\Delta}{2} - Bk^2\right)^2 + (\hbar v_F)^2 k^2}\right]^2 + (\hbar v_F)^2 k^2}.$$

(7.46)

The energy gap Δ and other model parameters are functions of thickness L and can be calculated numerically. The numerical results of Δ, v_F, D, and B are presented in Fig. 7.10. It is noted that $|D|$ must be less than $|B|$, otherwise the energy gap will disappear, and all discussions in the following will not be valid. The Δ terms play a role of mass term in $2 + 1$ Dirac equations.

In the large L limit,

$$v_F = \frac{A_{\parallel}}{\hbar} \sqrt{1 - \frac{D_{\perp}^2}{B_{\perp}^2}}. \tag{7.47}$$

The dispersion relation is given by

$$\varepsilon_{c/v}(k) = \pm v_F \hbar k \tag{7.48}$$

for small k. As a result, the energy gap closes at $k = 0$. The two massless Dirac cones are located near the top and bottom surfaces, respectively, as expected in a three-dimensional topological insulator.

In a small L limit,

$$v_F = \frac{A_{\parallel}}{\hbar}, \tag{7.49}$$

and

$$\Delta = \frac{2B_{\perp}\pi^2}{L^2}. \tag{7.50}$$

The ratio of the two velocities in the limits is

$$\eta = \frac{1}{\sqrt{1 - \frac{D_{\perp}^2}{B_{\perp}^2}}}. \tag{7.51}$$

It is noted that the velocity and energy gap for an ultrathin film are enhanced for a thinner film.

7.7.2 Structural Inversion Asymmetry

Recent experiments [47, 48] revealed that the substrate on which the film is grown influences dramatically electronic structure inside the film. Because the top surface of the film is usually exposed to the vacuum and the bottom surface is attached to a substrate, the inversion symmetry does not hold along z-direction, leading to the Rashba-like energy spectra for the gapped surface states. In this case, an extra term that describes the structure inversion asymmetry needs to be taken into account in the effective model.

Without loss of generality, we add a potential energy $V(z)$ into the Hamiltonian. Generally speaking, $V(z)$ can be expressed as $V(z) = V_s(z) + V_a(z)$, in which $V_s(z) = [V(z) + V(-z)]/2 = V_s(-z)$ and $V_a(z) = [V(z) - V(-z)]/2 = -V_a(-z)$. The symmetric term V_s could contribute to the mass term Δ in the effective model, which may lead to an energy splitting of the Dirac cone at the Γ point. Here we focus on the case of the antisymmetric term, $V(z) = V_a(z)$, which breaks the top-bottom inversion symmetry in the Hamiltonian. A detailed analysis gives the effective Hamiltonian for structure inversion asymmetry:

$$V_{\mathrm{eff}}^{\mathrm{SIA}} = \begin{bmatrix} 0 & 0 & \tilde{V} & 0 \\ 0 & 0 & 0 & \tilde{V}^* \\ \tilde{V}^* & 0 & 0 & 0 \\ 0 & \tilde{V} & 0 & 0 \end{bmatrix}, \tag{7.52}$$

where

$$\tilde{V} = \int_{-L/2}^{L/2} dz \langle \varphi^1 | V_a(z) | \chi^1 \rangle. \tag{7.53}$$

When the term of structure inversion asymmetry is included, the Hamiltonian in Eq. (7.42) with $V_{\mathrm{eff}}^{\mathrm{SIA}}$ Eq. (7.52) gives

$$E_{1,\pm} = E_0 - Dk^2 \pm \sqrt{\left(\frac{\Delta}{2} - Bk^2\right)^2 + (|\tilde{V}| + \hbar v_F k)^2}, \tag{7.54a}$$

$$E_{2,\pm} = E_0 - Dk^2 \pm \sqrt{\left(\frac{\Delta}{2} - Bk^2\right)^2 + (|\tilde{V}| - \hbar v_F k)^2}, \tag{7.54b}$$

where the extra index 1 (2) stands for the inner (outer) branches of the conduction or valence bands. Consequently, both the conduction and valence bands show Rashba-like splitting in the presence of structure inversion asymmetry. An intuitive understanding of the energy spectra can be given with the help of Fig. 7.11. On the left is for a thicker freestanding symmetric topological insulator film, and it has a single gapless Dirac cone on each of its two surfaces, with the solid and dash lines for the top and bottom surface, respectively. The two Dirac cones are degenerate. The top of Fig. 7.11 indicates that the inter-surface coupling across an ultrathin film will turn the Dirac cones into gapped Dirac hyperbolas. On the bottom of Fig. 7.11, the structure inversion asymmetry lifts the Dirac cone at the top surface while lowers the Dirac cone at the bottom surface. The potential difference at the top and bottom surfaces removes the degeneracy of the Dirac cones. On the right of Fig. 7.11, the coexistence of both the inter-surface coupling and structure inversion asymmetry leads to two gapped Dirac hyperbolas which also split in k-direction.

7.7.3 Experimental Data of ARPES

The thickness-dependent band structure of molecular beam epitaxy grown ultrathin film Bi$_2$Se$_3$ was investigated by angle-resolved photoemission spectroscopy by several groups [47, 48]. The energy gap due to the interlayer coupling has been observed experimentally in the surface states of ultrathin film Bi$_2$Se$_3$ below the thickness of 6QL. The spectrum splitting caused by structure inversion asymmetry was also confirmed. The observed experimental data can be fitted by the dispersion in Eq. (7.54) very well (Fig. 7.12).

Fig. 7.11 The evolution of
the doubly degenerate gapless
Dirac cones for the 2D
surface states, in the presence
of both inter-surface coupling
and structure inversion
asymmetry (*SIA*), into gapped
hyperbolas that also split in
the k-direction. The blue solid
and green dashed lines
correspond to the states
residing near the top and
bottom surfaces, respectively
(Adapted from [16])

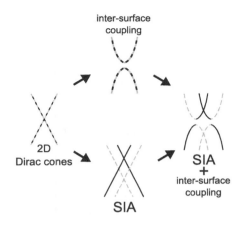

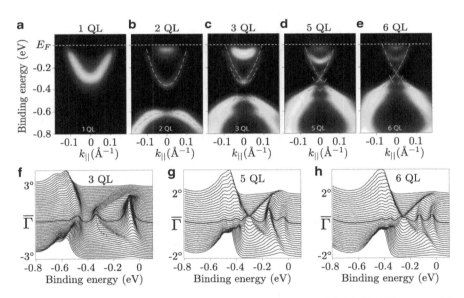

Fig. 7.12 ARPES spectra of Bi2Se3 thin film at room temperature. (**a**)–(**e**) ARPES spectra of 1,
2, 3, 4, 5, and 6 quintuple layer (*QL*) along $\Gamma - K$ direction. (**f**)–(**h**) Energy distribution curves of
(**c**), (**d**), and (**e**). The *pink dashed lines* in (b) represent the fitted curves using formula in Eq. (7.44).
The *blue* and *red dashed lines* in (**c**)–(**e**) represent the fitted curves using formula in Eq. (7.54)
(Adapted from [47])

The gap of the surface states is caused by the spatial confinement of thin film,
which does not break time reversal symmetry. This is different from the gap opening
of the surface states in a Zeeman field or magnetic impurity doping.

7.8 HgTe Thin Film

HgTe is a semimetal with an inverted band structure. Usually a strain will induce
an energy gap and force the HgTe to evolve into a topological insulator. However,
it is technically difficult to applying a strong strain on the sample to make a
semimetal insulating in experimental condition. The finite size effect provides a
practical way to open an energy gap in the bulk state when the dimensionality of the
sample is reduced from three dimensions to two dimensions as those in a quantum
well [14].

When the film is thin enough, the finite size caused band gap becomes obvious.
In this case, the finite confinement-induced sub-bands are far away from the low-
energy regime. We can then use the quantum well approximation $\langle k_z \rangle = 0$, $\langle k_z^2 \rangle \simeq
(\pi/L)^2$. Using these relations in the Hamiltonian in Eq. (6.62) and choosing the
basis set in the sequence $(|\psi_1\rangle, |\psi_3\rangle, |\psi_5\rangle, |\psi_2\rangle, |\psi_6\rangle, |\psi_4\rangle)$, we can obtain a two-
dimensional six-band Kane model:

$$H(\mathbf{k}) = \begin{pmatrix} h(\mathbf{k}) & 0 \\ 0 & h^*(-\mathbf{k}) \end{pmatrix}, \tag{7.55}$$

where

$$h(\mathbf{k}) = \begin{pmatrix} h_{11} & -\frac{1}{\sqrt{2}}Pk_+ & \frac{1}{\sqrt{6}}Pk_- \\ -\frac{1}{\sqrt{2}}Pk_- & h_{22} & \sqrt{3}\overline{\gamma}Bk_-^2 \\ \frac{1}{\sqrt{6}}Pk_+ & \sqrt{3}\overline{\gamma}Bk_+^2 & h_{33} \end{pmatrix} \tag{7.56}$$

with

$$h_{11} = E_g + B(2F + 1)(k_\parallel^2 + \langle k_z^2 \rangle), \tag{7.57}$$

$$h_{22} = -(\gamma_1 + \overline{\gamma})Bk_\parallel^2 - (\gamma_1 - 2\overline{\gamma})B\langle k_z^2 \rangle, \tag{7.58}$$

$$h_{33} = -(\gamma_1 - \overline{\gamma})Bk_\parallel^2 - (\gamma_1 + 2\overline{\gamma})B\langle k_z^2 \rangle. \tag{7.59}$$

The system keeps time reversal symmetry, and the representation of the symmetry
operation in the new set of basis is given by $T = \mathcal{K} \cdot i\sigma^y \otimes I_{3\times3}$, where $\mathcal{K}$ is
the complex conjugation operator, σ^y, and I denote the Pauli matrix and identity
matrix, respectively.

We can study the two blocks separately since they are time reversal counterparts
of each other. Here we focus on the upper block first. At $k_x = 0$, the boundaries of
Γ_6, light hole (LH), and heavy hole (HH) are at $E = E_g + B(2F + 1)\langle k_z^2 \rangle$, $E =
-(\gamma_1 - 2\overline{\gamma})B\langle k_z^2 \rangle$, and $E = -(\gamma_1 + 2\overline{\gamma})B\langle k_z^2 \rangle$, which are controllable by choosing
film thickness L. Down to $L \approx 30$ Å, Γ_6 band flips up and exchanges position
with HH, the system is still nontrivial. Further down to $L \approx 20$ Å, Γ_6 flips up and
exchanges with the conduction band. The band structure becomes trivial. Using the
tight binding approximation, we can transform $h(\mathbf{k})$ into a tight binding model on a
two-dimensional lattice. In Fig. 7.13, we show the local density of states on the edge

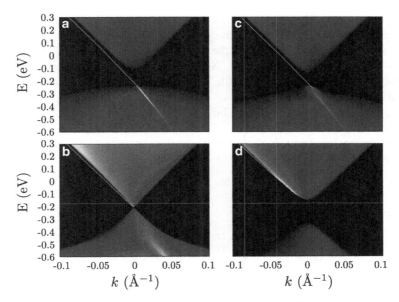

Fig. 7.13 Local density of states at the edge of thin films at different thickness using the Hamiltonian in Eq. (7.56) with 2D lattice model in the tight binding approximation. (**a**) L = 28A; (**b**) L = 25A; (**c**) L = 20A; (**d**) L = 16A (Adapted from [14])

a semi-infinite film for $h(\mathbf{k})$. When $L > 20$ Å, the edge states are found connecting the valence and conduction bands. After the system becomes trivial when $L < 20$ Å, the edge states do not cross the band gap anymore; instead they only attach to the valence band. At the critical point $L = 20$ Å, the valence band and conduction band touch and form a linear Dirac cone at the low-energy regime. This shows that by controlling the film thickness, it is possible to obtain a single-valley Dirac cone for each spin block without using the topological surface states [49]. Notice that in Fig. 7.13, we can also see the edge states submerging in the valence bands.

7.9 Further Reading

- L. Fu, C.L. Kane, E.J. Mele, Topological insulators in three dimensions. Phys. Rev. Lett. **98**, 106803 (2007)
- J.E. Moore, L. Balents, Topological invariants of time reversal-invariant band structures. Phys. Rev. B **75**, 121306(R) (2007)
- R. Roy, Topological phases and the quantum spin Hall effect in three dimensions. Phys. Rev. B **79**, 195322 (2009)
- D. Hsieh, D. Qian, L. Wray, Y. Xia, Y.S. Hor, R.J. Cava, M.Z. Hasan, A topological Dirac insulator in a quantum spin Hall phase. Nature **452**, 970 (2008)

- Y. Xia, D. Qian, D. Hsieh, L. Wray, A. Pal, H. Lin, A. Bansil, D. Grauer, Y.S. Hor, R.J. Cava, M.Z. Hasan, Observation of a large-gap topological-insulator class with a single Dirac cone on the surface. Nat. Phys. **5**, 398 (2009)
- H.J. Zhang, C.X. Liu, X.L. Qi, X. Dai, Z. Fang, S.C. Zhang, Topological insulators in Bi_2Se_3, Bi_2Te_3 and Sb_2Te_3 with a single Dirac cone on the surface. Nat. Phys. **5**, 438 (2009)
- Y.L. Chen, J.G. Analytis, J.-H. Chu, Z.K. Liu, S.-K. Mo, X.L. Qi, H.J. Zhang, D.H. Lu, X. Dai, Z. Fang, S.C. Zhang, I.R. Fisher, Z. Hussain, Z.X. Shen, Experimental realization of a three-dimensional topological insulator, Bi_2Te_3. Science **325**, 178 (2009)

References

1. W.W. Scanlon, Solid State Phys. **9**, 83 (1959)
2. J.O. Dimmock, I. Melngailis, A.J. Strauss, Phys. Rev. Lett. **16**, 1193 (1966)
3. J.O. Dimmock, G.B. Wright, Phys. Rev. **135**, A821 (1964)
4. J.B. Conklin, Jr., L.E. Johnson, G.W. Pratt, Jr., Phys. Rev. **137**, A1282 (1965)
5. M.Z. Hasan, C.L. Kane, Rev. Mod. Phys. **82**, 3045 (2010)
6. L. Fu, C.L. Kane, Phys. Rev. B **76**, 045302 (2007)
7. D. Hsieh, D. Qian, L. Wray, Y. Xia, Y.S. Hor, R.J. Cava, M.Z. Hasan, Nature (London) **452**, 970 (2008)
8. L.S. Lerner, K.F. Cuff, L.R. Williams, Rev. Mod. Phys. **40**, 770 (1968)
9. B. Lenoir, M. Cassart, J.P. Michenaud, H. Scherrer, S. Scherrer, J. Phys. Chem. Solids **57**, 89 (1996)
10. Y. Xia, D. Qian, D. Hsieh, L. Wray, A. Pal, H. Lin, A. Bansil, D. Grauer, Y.S. Hor, R.J. Cava, M.Z. Hasan, Nat. Phys. **5**, 398 (2009)
11. H. Zhang, C.X. Liu, X.L. Qi, X. Dai, Z. Fang, S.C. Zhang, Nat. Phys. **5**, 438 (2009)
12. Y.L. Chen, J.G. Analytis, J.H. Chu, Z.K. Liu, S.K. Mo, X.L. Qi, H.J. Zhang, D.H. Lu, X. Dai, Z. Fang, S.C. Zhang, I.R. Fisher, Z. Hussain, Z.X. Shen, Science **325**, 178 (2009)
13. C. Brüne, C.X. Liu, E.G. Novik, E.M. Hankiewicz, H. Buhmann, Y.L. Chen, X.L. Qi, Z.X. Shen, S.C. Zhang, L.W. Molenkamp, Phys. Rev. Lett. **106**, 126803 (2011)
14. R.L. Chu, W.Y. Shan, J. Lu, S.Q. Shen, Phys. Rev. B **83**, 075110 (2011)
15. W.X. Feng, D. Xiao, J. Ding, Y.G. Yao, Phys. Rev. Lett. **106**, 016402 (2011)
16. W.Y. Shan, H.Z. Lu, S.Q. Shen, New J. Phys. **12**, 043048 (2010)
17. L. Fu, Phys. Rev. Lett. **103**, 266801 (2009)
18. P. Roushan, J. Seo, C.V. Parker, Y.S. Hor, D. Hsieh, D. Qian, A. Richardella, M.Z. Hasan, R.J. Cava, A. Yazdani, Nature **460**, 1106 (2009)
19. T. Zhang, P. Cheng, X. Chen, J.F. Jia, X.C. Ma, K. He, L.L. Wang, H.J. Zhang, X. Dai, Z. Fang, X.C. Xie, Q.K. Xue, Phys. Rev. Lett. **103**, 266803 (2009)
20. J.G. Checkelsky, Y.S. Hor, M.-H. Liu, D.-X. Qu, R.J. Cava, N.P. Ong, Phys. Rev. Lett. **103**, 246601 (2009)
21. H. Peng, K. Lai, D. Kong, S. Meister, Y. Chen, X.L. Qi, S.C. Zhang, Z.X. Shen, Y. Cui, Nat. Mater. **9**, 225 (2010)
22. J.G. Checkelsky, Y.S. Hor, R.J. Cava, N.P. Ong, Phys. Rev. Lett. **106**, 196801 (2011)
23. J. Chen, H.J. Qin, F. Yang, J. Liu, T. Guan, F.M. Qu, G.H. Zhang, J.R. Shi, X.C. Xie, C.L. Yang, K.H. Wu, Y.Q. Li, L. Lu, Phys. Rev. Lett. **105**, 176602 (2010)
24. H.T. He, G. Wang, T. Zhang, I.K. Sou, G.K.L. Wong, J.N. Wang, H.Z. Lu, S.Q. Shen, F.C. Zhang, Phys. Rev. Lett. **106**, 166805 (2011)
25. T. Ando, T. Nakanishi, R. Saito, J. Phys. Soc. J. **67**, 2857 (1998)

26. D. Hsieh, Y. Xia, D. Qian, L. Wray, J.H. Dil, F. Meier, J. Osterwalder, L. Patthey, J.G. Checkelsky, N.P. Ong, A.V. Fedorov, H. Lin, A. Bansil, D. Grauer, Y.S. Hor, R.J. Cava, M.Z. Hasan, Nature (London) **460**, 1101 (2009)
27. A.A. Taskin, Y. Ando, Phys. Rev. B **80**, 085303 (2009)
28. A.A. Taskin, K. Segawa, Y. Ando, Phys. Rev. B **82**, 121302(R) (2010)
29. K. Eto, Z. Ren, A.A. Taskin, K. Segawa, Y. Ando, Phys. Rev. B **81**, 195309 (2010)
30. N.P. Butch, K. Kirshenbaum, P. Syers, A.B. Sushkov, G.S. Jenkins, H.D. Drew, J. Paglione, Phys. Rev. B **81**, 241301(R) (2010)
31. J.G. Analytis, J.H. Chu, Y.L. Chen, F. Corredor, R.D. McDonald, Z.X. Shen, I.R. Fisher, Phys. Rev. B **81**, 205407 (2010)
32. G.P. Mikitik, Y.V. Sharlai, Phys. Rev. Lett. **82**, 2147 (1999)
33. X.L. Qi, T.L. Hughes, S.C. Zhang, Phys. Rev. B **78**, 195424 (2008)
34. B.I. Halperin, Phys. Rev. B **25**, 2185 (1982)
35. A.H. MacDonald, P. Streda, Phys. Rev. B **29**, 1616 (1984)
36. A.N. Redlich, Phys. Rev. D **29**, 2366 (1984)
37. R. Jackiw, Phys. Rev. D **29**, 2375 (1984)
38. H.Z. Lu, W.Y. Shan, W. Yao, Q. Niu, S.Q. Shen, Phys. Rev. B **81**, 115407 (2010)
39. M. Büttiker, Phys. Rev. Lett. **57**, 1761 (1986)
40. S. Datta, *Electronic Transport in Mesoscopic Systems* (Cambridge University Press, Cambridge, 1995)
41. J. Li, L.B. Hu, S.Q. Shen, Phys. Rev. B **71**, 241305(R) (2005)
42. R.L. Chu, J.R. Shi, S.Q. Shen, Phys. Rev. B **84**, 085312 (2011)
43. S.Q. Shen, M. Ma, X.C. Xie, F.C. Zhang, Phys. Rev. Lett. **92**, 256603 (2004)
44. K.S. Novoselov, Z. Jiang, Y. Zhang, S.V. Morozov, H.L. Stormer, U. Zeitler, J.C. Maan, G.S. Boebinger, P. Kim, A.K. Geim, Science **315**, 1379 (2007)
45. P. Cheng, C.L. Song, T. Zhang, Y.Y. Zhang, Y.L. Wang, J.F. Jia, J. Wang, Y.Y. Wang, B.F. Zhu, X. Chen, X.C. Ma, K. He, L.L. Wang, X. Dai, Z. Fang, X.C. Xie, X.L. Qi, C.X. Liu, S.C. Zhang, Q.K. Xue, Phys. Rev. Lett. **105**, 076801 (2010)
46. T. Hanaguri, K. Igarashi, M. Kawamura, H. Takagi, T. Sasagawa, Phys. Rev. B **82**, 081305(R) (2010)
47. Y. Zhang, K. He, C.Z. Chang, C.L. Song, L.L. Wang, X. Chen, J.F. Jia, Z. Fang, X. Dai, W.Y. Shan, S.Q. Shen, Q. Niu, X.L. Qi, S.C. Zhang, X.C. Ma, Q.K. Xue, Nat. Phys. **6**, 584 (2010)
48. Y. Sakamoto, T. Hirahara, H. Miyazaki, S. Kimura, S. Hasegawa, Phys. Rev. B **81**, 165432 (2010)
49. B. Büttner, C.X. Liu, G. Tkachov, E.G. Novik, C. Brüne, H. Buhmann, E.M. Hankiewicz, P. Recher, B. Trauzettel, S.C. Zhang, L.W. Molenkamp, Nat. Phys. **7**, 418 (2011)

Chapter 8
Impurities and Defects in Topological Insulators

Abstract Impurities and defects in topological insulators can be regarded as a boundary of the system. The bound states may be formed around these impurities or defects for the same reason as the formation of the edge or surface states.

Keywords In-gap bound state • Topological defects • Wormhole effect • Witten effect

Topological insulators are distinguished from conventional band insulators according to the Z_2 invariant classification of the band insulators that respect time reversal symmetry. Variation of the Z_2 invariants at their boundaries will lead to the topologically protected edge or surface states with the gapless Dirac energy spectrum. Impurities or defects are inevitably present in topological insulators. They may change the geometry or topology of the systems and induce the bound states as those near the boundary. Reminding that the boundary state is a manifestation of the topological nature of bulk bands, one should start with the examination of the host bulk to know how impurities or defects affect the electronic structure. It was known that a single impurity or defect can induce bound states in many systems, such as in the Yu-Shiba state in s-wave superconductor [1, 2] and in d-wave superconductors [3]. In this chapter, we study that bound states around a single vacancy or defect in the bulk energy gap of topological insulators.

8.1 One Dimension

When a δ potential $V(x) = V_0\delta(x)$ is present in an infinite one-dimensional topological insulator, the equation for the wave function reads

$$\left[vp_x\sigma_x + \left(mv^2 - Bp_x^2\right)\sigma_z + V_0\delta(x)\right]\Psi(x) = E\Psi(x), \tag{8.1}$$

S.-Q. Shen, *Topological Insulators: Dirac Equation in Condensed Matters*,
Springer Series in Solid-State Sciences 174, DOI 10.1007/978-3-642-32858-9_8,
© Springer-Verlag Berlin Heidelberg 2012

where $\Psi(x)$ is a two-component spinor. The continuity of the wave function at $x = 0$ requires

$$\lim_{\epsilon \to 0^+} \Psi(\epsilon) = \Psi(-\epsilon). \tag{8.2}$$

In addition, the integral of Eq. (8.1) around the δ potential leads to

$$\lim_{\epsilon \to 0^+} [\partial_x \Psi|_{x=\epsilon} - \partial_x \Psi|_{x=-\epsilon}] = -\frac{V_0}{B\hbar^2} \sigma_z \Psi(0), \tag{8.3}$$

that is, the derivative of the wave function is not continuous at $x = 0$. To find a bound state near $x = 0$, the electron wave function should vanish when $x \to \pm\infty$.
For $x > 0$,

$$\Psi(x > 0) = c_1^+ e^{-x/\xi_1} + c_2^+ e^{-x/\xi_2}, \tag{8.4}$$

and for $x < 0$,

$$\Psi(x < 0) = c_1^- e^{+x/\xi_1} + c_2^- e^{+x/\xi_2}, \tag{8.5}$$

with $\xi_{1,2}^{-1} = \frac{|v|}{2|B|\hbar}(1 \pm \sqrt{1 - 4mB})$.
Substituting the wave function into Eqs. (8.2) and (8.3) at $x = 0$, one obtains two transcendental equations for the bound state energy

$$\sqrt{1 - 2mB + 2|mB|\sqrt{1 - \frac{E^2}{m^2 v^4}}} = \frac{V_0}{2\hbar v} \left[\frac{\pm mv^2 - E}{\sqrt{m^2 v^4 - E^2}} \mp \mathrm{sgn}(B) \right], \tag{8.6}$$

where up to two solutions can be found. When $V_0 = 0$ and the δ potential vanishes, there is no solution to equation which satisfies the boundary condition.

The formation of the bound states essentially has the same origin as the boundary states in topological insulators (Fig. 8.1). Consider an infinite one-dimensional topological insulator, in which the energy gap separates the positive and negative spectra. If we cut the chain at one point, saying $x = 0$, then we produce two open boundaries at the two sides of $x > 0$ and $x < 0$. There exists a pair of states (end states) at the boundaries with the same energy

$$\Psi(\pm) = \frac{C}{\sqrt{2}} \begin{pmatrix} \pm\mathrm{sgn}(B) \\ i \end{pmatrix} (e^{\mp x/\xi_1} - e^{\mp x/\xi_2}), \tag{8.7}$$

with $C = \sqrt{2(\xi_1 + \xi_2)}/|\xi_1 - \xi_2|$. " $\pm$ " indicates that the semi-infinite chain lies in the region $x > 0$ or < 0. The energies of these states lie inside the bulk gap, and are equal to 0. Now we paste the two ends again with some kind of "glue potential"; it is possible that these end states can be trapped or mixed around the connecting point and evolve into in-gap bound states. The shapes of the possibility density of the wave function of our solutions for a δ−potential support this intuitive picture for the formation of the in-gap bound states. An impurity locating at $x = 0$, unlike the

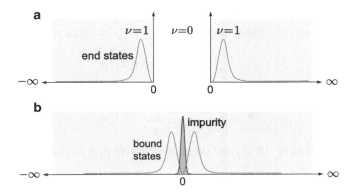

Fig. 8.1 In-gap bound states in one-dimensional topological insulator. (**a**) The presence of end states with zero energy at the open boundaries of a broken one-dimensional topological insulator. (**b**) The zero-energy end states evolve into in-gap bound states when the two open boundaries are connected by an impurity (Adapted from [4])

open boundary, allows tunneling between the two ends of the chain and will affect the behavior of the wave function near the point of $x = 0$. For a δ−potential, the bound states induced by it are always there regardless of its strength.

For comparison, consider an ordinary insulator of $mB < 0$. A pair of bound states induced by a δ−potential is also possible when $0 < |V_0| < 2\hbar|v|\sqrt{1 - 2mB}$, but vanishes after $|V_0|$ exceeds $2\hbar|v|\sqrt{1 - 2mB}$, indicating the distinct origin from those for $mB > 0$.

8.2 Integral Equation for Bound State Energies

The bound states can be formally obtained by solving an integral equation. Although in most cases the integral equation cannot be solved analytically, it does provide rich information about the existence of bound states under some certain impurity potentials in various dimensions. The modified Dirac equation with a potential $V(\mathbf{r})$ can be written as

$$[E - H_0(\mathbf{r})]\Psi(\mathbf{r}) = V(\mathbf{r})\Psi(\mathbf{r}). \tag{8.8}$$

The wave function $\Psi(\mathbf{r})$ can be expanded by its Fourier transformation components as

$$\Psi(\mathbf{r}) = \sum_{\mathbf{p}'} u_{\mathbf{p}'} e^{i\mathbf{p}' \cdot \mathbf{r}/\hbar}. \tag{8.9}$$

Thus, one obtains

$$[E - H_0(\mathbf{p})]u_{\mathbf{p}} = \sum_{\mathbf{p}'} V_{\mathbf{p}\mathbf{p}'} u_{\mathbf{p}'}, \tag{8.10}$$

where $V_{\mathbf{pp}'} = \int d\mathbf{r}V(\mathbf{r})e^{-i(\mathbf{p}-\mathbf{p}')\cdot\mathbf{r}/\hbar}$. While this equation cannot be solved analytically in general, one can find the solution if $V_{\mathbf{pp}'}$ is taken to be a factorizable potential [5]

$$V_{\mathbf{pp}'} = V_0\xi^*(\mathbf{p})\xi(\mathbf{p}'). \tag{8.11}$$

In this case,

$$u_{\mathbf{p}} = \frac{V_0\xi^*(\mathbf{p})}{E - H_0(\mathbf{p})} \sum_{\mathbf{p}'} \xi(\mathbf{p}')u_{\mathbf{p}'}. \tag{8.12}$$

Multiplying $\xi(p)$ in Eq. (8.12) and summarizing over p, it follows that

$$\left[\sum_{\mathbf{p}} \frac{V_0\xi^*(\mathbf{p})\xi(\mathbf{p})}{E - H_0(\mathbf{p})} - 1\right] \sum_{\mathbf{p}'} \xi(\mathbf{p}')u_{\mathbf{p}'} = 0. \tag{8.13}$$

Thus, one obtains

$$\det\left[\sum_{\mathbf{p}} \frac{V_0\xi^*(\mathbf{p})\xi(\mathbf{p})}{E - H_0(\mathbf{p})} - 1\right] = 0. \tag{8.14}$$

For a magnetic impurity, it will be more complicated.

More generally, if the system is isotropic and $V_{\mathbf{pp}'}$ can be expanded into its partial wave components

$$V_{\mathbf{pp}'} = \sum_{l=0}^{\infty} \sum_{m=-l}^{l} V(|\mathbf{p}|, |\mathbf{p}'|)Y_l^m(\Omega_{\mathbf{p}})Y_l^{-m}(\Omega_{\mathbf{p}'}) \tag{8.15}$$

with a factorizable $V(|\mathbf{p}|, |\mathbf{p}'|) = \lambda_l w_{\mathbf{p}}^l(w_{\mathbf{p}'}^l)^*$ where $Y_l^m(\Omega_{\mathbf{p}})$ is the spherical harmonic Bessel function, the solution can be determined.

8.2.1 δ–potential

For a delta potential $V(\mathbf{r}) = V_0\delta(\mathbf{r})$, $V_{\mathbf{pp}'} \equiv V_0$. A nontrivial solution requires

$$\det\left[\sum_{\mathbf{p}} \frac{V_0}{E - H_0(\mathbf{p})} - 1\right] = 0, \tag{8.16}$$

or

$$\det\left[\int \frac{d^d\mathbf{p}}{(2\pi\hbar)^d} \frac{V_0}{E - H_0(\mathbf{p})} - 1\right] = 0, \tag{8.17}$$

where d is the dimensionality.

For the one-dimensional case, the modified Dirac Hamiltonian can be easily inverted. After some algebra, we have

$$\int_0^{+\infty} \frac{dk_x}{\pi} \frac{[E_{A/B} \pm (mv^2 - B\hbar^2 k_x^2)]V_0}{E_{A/B}^2 - (mv^2 - B\hbar^2 k_x^2)^2 - v^2\hbar^2 k_x^2} = 1, \tag{8.18}$$

where E_A and E_B denote the energy solution for "+" and "−," respectively. From this equation we can recover the result in Sect. 8.1.

For the two-dimensional case, one can obtain a similar integral equation for the two-dimensional bound state energies,

$$\int_0^{+\infty} \frac{kdk}{2\pi} \frac{[E_{A/B} \pm (mv^2 - B\hbar^2 k^2)]V_0}{E_{A/B}^2 - (mv^2 - B\hbar^2 k^2)^2 - v^2\hbar^2 k^2} = 1, \tag{8.19}$$

where $k^2 = k_x^2 + k_y^2$. However, the integral in Eq. (8.19) will logarithmically diverge when $|k| \to +\infty$. This means in two-dimensional, an impurity with δ-potential cannot trap any bound states. Similarly in three dimensions, although the integration equation is more complicated, divergence also exists in the k-integration, which excludes the possibility of three-dimensional bound states under δ-potential.

Considering the Brillouin zone of lattice crystal is always finite, it is possible to form bound states under δ−potential by introducing a reasonable cutoff of k.

8.3 Bound States in Two Dimensions

The formation of the in-gap bound states can be readily illustrated by reviewing the edge states in two-dimensional topological insulators. As the Z_2 index varies across the boundary, the edge states arise in the gap with the gapless Dirac dispersion. Unlike the quantum Hall effect in a magnetic field, spin-orbit coupling respects time reversal symmetry, so the resulting edge states appear in pairs, of which one state is the time reversal counterpart of the other, propagating in opposite directions and with opposite spins (Fig. 8.2b). Now imagine that the system edge is bent into a hole or punch a large hole in the system, the edge states will circulate around the hole as the periodic boundary conditions along the propagating direction remain unchanged (Fig. 8.2d). The dispersion of this edge state is proportional to $\left(n + \frac{1}{2}\right) \hbar/R$ ($n\hbar$ is for orbital angular momentum). While shrinking the radius of the hole, most of the edge states will be expelled into the bulk bands as the energy separation between the states becomes larger and larger for smaller R. It is found that at least two degenerate pairs of the states will be trapped to form the bound states in the gap as the hole shrinks into a point defect. This mechanism of the formation of the bound states can be realized in topological insulator in all dimensions.

In two dimensions, the modified Dirac model can be reduced into two independent 2×2 Hamiltonians

$$h_\pm = (mv^2 - B\mathbf{p}^2)\sigma_z + v(\mathbf{p}_x\sigma_x \pm \mathbf{p}_y\sigma_y), \tag{8.20}$$

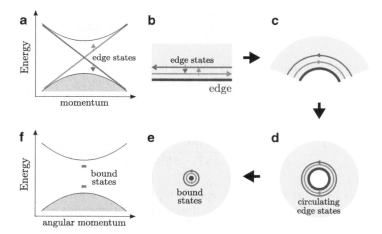

Fig. 8.2 Schematic description of the formation of vacancy-induced in-gap bound states in two-dimensional topological insulators. (**a**) and (**b**) A pair of helical edge states traveling along the edge of a two-dimensional topological insulator with the gapless Dirac dispersion. (**c**) and (**d**) When the edge is bent into a hole, the helical edge states evolve to circulate around the hole. (**e**) and (**f**) The circulating edge states may develop into bound states as the hole shrinks into a point or being replaced by a vacancy (Adapted from [6])

with h_- the time reversal counterpart of h_+ [7–9]. It is convenient to adopt the polar coordinates $(x, y) = r(\cos\varphi, \sin\varphi)$ in two dimensions. In the coordinate

$$\mathbf{p}_\pm = -i\hbar e^{\pm i\theta}\left(\partial_r \pm \frac{i\partial_\theta}{r}\right), \tag{8.21a}$$

$$\mathbf{p}^2 = -\hbar^2\left(\partial_r^2 + \frac{1}{r}\partial_r + \frac{1}{r^2}\partial_\theta^2\right). \tag{8.21b}$$

Here these equations are solved under the Dirichlet boundary conditions (Fig. 8.3a), that is, the center of the two-dimensional topological insulator is punched with a hole of radius R; thus, the wave function is required to vanish at $r = R$ and $r = +\infty$. Due to the rotational symmetry of h_+, it is found that the z-component of the total angular momentum $J_z = -i\hbar\partial_\theta + (\hbar/2)\sigma_z$ satisfies

$$[h_+, J_z] = 0 \tag{8.22}$$

and provides a good quantum number. The wave function has a general form

$$\varphi_l(r, \theta) = \begin{pmatrix} f_l(r)e^{il\theta} \\ g_l(r)e^{i(l+1)\theta} \end{pmatrix} \tag{8.23}$$

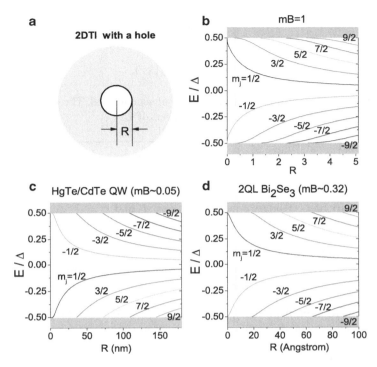

Fig. 8.3 Two-dimensional in-gap bound states. (**a**) A two-dimensional topological insulator with a hole of radius R at the center. (**b**) and (**c**) Energies (E in units of the band gap Δ) of in-gap bound states circulating around the hole as functions of the hole radius. m_j is the quantum number for the z-component of the total angular momentum of the circulating bound states. In (**b**), $m=v=B=\hbar=1$; in (**c**), $mv^2 = -10\,\text{meV}$, $B\hbar^2 = -686\,\text{meV·nm}^2$, and $\hbar v = 364.5\,\text{meV·nm}$ (Adopted from Ref. [7]); in (**d**), $mv^2 = 0.126\,\text{eV}$, $B\hbar^2 = 21.8\,\text{eV·Å}^2$, $\hbar v = 2.94\,\text{eV·Å}$(Adopted from Ref. [11]). $\Delta = 2mv^2$ for $0 < mB < 1/2$, and $\Delta = (v^2/|B|)\sqrt{4mB - 1}$ for $mB > 1/2$. The *gray* areas line mark the bulk bands (Adapted from [6])

with an integer l, satisfying that $J_z\varphi_l(r, \theta) = j\hbar\varphi_l(r, \theta)$ with $j = l + \frac{1}{2}$. Thus, the equation is reduced that for the radial part of the wave function,

$$h_{\text{eff}}\begin{pmatrix} f_l(r) \\ g_l(r) \end{pmatrix} = E\begin{pmatrix} f_l(r) \\ g_l(r) \end{pmatrix}, \tag{8.24}$$

where

$$h_{\text{eff}} = \begin{pmatrix} mv^2 + \hbar^2 B(\partial_r^2 + \frac{\partial_r}{r} - \frac{l^2}{r^2}) & -i\hbar v(\partial_r + \frac{l+1}{r}) \\ -i\hbar v(\partial_r - \frac{l}{r}) & -mv^2 - \hbar^2 B(\partial_r^2 + \frac{\partial_r}{r} - \frac{(l+1)^2}{r^2}) \end{pmatrix}. \tag{8.25}$$

We take the trial wave function

$$f_l(r) = c_l K_l(\lambda r),$$ (8.26a)

$$g_l(r) = d_l K_{l+1}(\lambda r),$$ (8.26b)

where $K_n(x)$ is the modified Bessel function of second kind. The secular equations give four roots of λ_n $(= \pm\lambda_1, \pm\lambda_2)$ as functions of E:

$$\lambda_{1,2}^2 = \frac{v^2}{2B^2\hbar^2}\left[1 - 2mB \pm \sqrt{1 - 4mB + \frac{4B^2E^2}{v^4}}\right].$$ (8.27)

Using the Dirichlet boundary conditions at $r = R$ and $r = +\infty$, we arrive at the transcendental equation for the bound-state energies

$$\frac{\lambda_1^2 + \dfrac{mv^2 - E}{B\hbar^2}}{\lambda_1}\frac{K_{l+1}(\lambda_1 R)}{K_l(\lambda_1 R)} = \frac{\lambda_2^2 + \dfrac{mv^2 - E}{B\hbar^2}}{\lambda_2}\frac{K_{l+1}(\lambda_2 R)}{K_l(\lambda_2 R)}.$$ (8.28)

Since there are more than one value of λ, the wave function should be the linear combination of the modified Bessel functions, for example, $f_l(r) = c_1 K_l(\lambda_1 r) + c_2 K_l(\lambda_2 r)$. With the boundary condition at $r = R$, the wave function $\varphi_l(r, \theta)$ for h_+ turns out to have the form

$$\begin{bmatrix} \dfrac{K_l(\lambda_1 R)}{K_{l+1}(\lambda_1 R)} f_l(r) e^{il\theta} \\[2em] i\dfrac{\lambda_1^2 + \dfrac{mv^2 - E}{B\hbar^2}}{(\lambda_1 v/B\hbar)} f_{l+1}(r) e^{i(l+1)\theta} \end{bmatrix}$$ (8.29)

with

$$f_l(r) = \frac{K_l(\lambda_1 r)}{K_l(\lambda_1 R)} - \frac{K_l(\lambda_2 r)}{K_l(\lambda_2 R)}.$$ (8.30)

The solution for h_- can be derived following the same procedure.

In Fig. 8.3c, d, we show the bound-state energies as functions of R for an ideal case (Fig. 8.3b, $mB = 1$), for the HgTe quantum well (Fig. 8.3c, $mB = 0.05$), and for a two-quintuple layer of Bi_2Se_3 thin film (Fig. 8.3d, $mB = 0.32$). For a macroscopically large R, we found an approximated solution for the energy spectrum of h_+ as $E = \left(l + \frac{1}{2}\right)\hbar v \operatorname{sgn}(B)/R$. As the time reversal copy of h_+, h_- has an approximated spectrum $E = -\left(l + \frac{1}{2}\right)\hbar v \operatorname{sgn}(B)/R$. They form a series of paired helical edge states, in good agreement with the edge-state solutions in the two-dimensional quantum spin Hall system [10] if we take $k = \left(l + \frac{1}{2}\right)/R$ for a large R. When shrinking R, the energy separation of these edge state $\Delta E = \hbar v /R$ increases with shrunk R, and the edge states with higher l will be pushed out of the energy gap gradually. However, we observe that for $mB > 0$, the state with $l = 0$ always stay in the energy gap, and as $R \rightarrow 0$, their energies approach to $E = \pm(v^2/2|B|)\sqrt{4mB - 1}$ for $mB > \frac{1}{2}$ or $\pm mv^2$ for $0 < mB < \frac{1}{2}$. When comparing the details of Fig. 8.3c with d, we find that the two pairs of states for $l = 0$ have quite different asymptotic behaviors in the spectrum when R decreases to zero. This can be explained by noting the fact that there is no in-gap bound state when $mB < 0$, suggesting $mB = 0$ is the critical point for the topological phase transition. The bound state with smaller mB is closer to the transition point and thus tends to enter the bulk more easily.

The solutions verify the formation of the in-gap bound states as shown in Fig. 8.2. Therefore, considering the symmetry between h_+ and h_-, we conclude that in the presence of vacancy or defect, there always exist *at least* two pairs of bound states in the energy gap in the two-dimensional quantum spin Hall system.

8.4 Topological Defects

There are several types of topological defects, such as magnetic monopoles, vortex line, or domain walls. In this book, we have already solved the problem of domain wall with a kind of mass distribution in Sect. 2.2. The solution of zero energy is quite robust against the distribution of domain wall. The solution has a lot of applications in polymers. The charge and spin carriers in one-dimensional polyacetylene are topological excitations generated by the domain wall. Here we present a solution of zero-energy mode for a quantum vortex in the quantum Hall system and its application to three-dimensional system.

8.4.1 Magnetic Flux and Zero-Energy Mode

When a magnetic flux is threading the hole, the energy levels of the in-gap bound states can be continuously manipulated (Fig. 8.4). Consider a magnetic flux ϕ that threads through the hole of a radius R. We perform the Peierls substitution $\mathbf{p} \rightarrow \mathbf{p} + e\mathbf{A}$ in h_+ in Eq. (8.20) by taking the gauge $\mathbf{A} = (\Phi/2\pi r)\mathbf{e}_\theta$, which still keeps m_j a good quantum number. Therefore, the eigenfunctions of this new Hamiltonian can be readily expressed as $\exp(-i\nu\theta)\varphi_l(r, \theta)$ after a gauge transformation, with $\nu = \phi/\phi_0$ and the flux quantum $\phi_0 = h/e$. In this case, the allowed value for the total angular momentum becomes $j = l + \frac{1}{2} + \nu$. The energies of in-gap bound states vary with the radius of the hole and the magnetic flux. When $\nu = \frac{1}{2}$ or $-\frac{1}{2}$, there always exists one solution of $j = 0$. In this case, the solution has a general form of

$$\varphi_{j=0}(r, \theta) = \begin{pmatrix} f_{\frac{1}{2}}(r)e^{-i\frac{\theta}{2}} \\ g_{\frac{1}{2}}(r)e^{+i\frac{\theta}{2}} \end{pmatrix}. \tag{8.31}$$

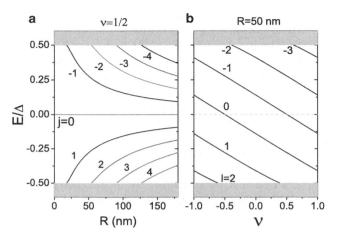

Fig. 8.4 Effect of magnetic flux on in-gap bound states. Energies (E in units of the band gap Δ) of in-gap bound states circulating around the hole as functions of (**a**) the hole radius R when half-quantum flux $\nu = 1/2$ is applied and (**b**) the magnetic flux ν (in unit of flux quantum $\Phi_0 = h/e$) for fixed radius $R = 50$ nm. m_+ (m_-) is the quantum number for the z-component of the total angular momentum j_{z+} (j_{z-}) of the circulating bound states. $m_j = m_+ + \nu$. In (**b**), *black* (*light gray*) *lines* belong to h_+ (h_-) block. In (**a**) and (**b**), $mv^2 = -10$ meV, $B\hbar^2 = -686$ meV·nm^2, and $\hbar v = 364.5$ meV·nm (Adopted from Ref. [7]). $\Delta = 2mv^2$. The *gray* areas mark the bulk bands

Equivalently, we replace l in Eq. (8.25) by $-\frac{1}{2}$:

$$\left[E + i\hbar v(\partial_r + \frac{1}{2r})\sigma_x - \left[mv^2 + \hbar^2 B\left(\partial_r^2 + \frac{\partial_r}{r} - \frac{1}{4r^2}\right)\right]\sigma_z\right]\begin{pmatrix} f \\ g \end{pmatrix} = 0. \tag{8.32}$$

Under a transformation,

$$\begin{pmatrix} f_{\frac{1}{2}}(r) \\ g_{\frac{1}{2}}(r) \end{pmatrix} = \frac{1}{\sqrt{r}}\phi(r), \tag{8.33}$$

the equation for the radial part of the wave function is reduced into a one-dimensional modified Dirac equation:

$$\left[-i\hbar v\partial_r\sigma_x + \left(mv^2 + \hbar^2 B\partial_r^2\right)\sigma_z\right]\phi(r) = E\phi(r). \tag{8.34}$$

There exists one bound state solution with zero energy near the boundary as we obtained in Sect. 2.5.1. As a result, it is found that

$$\varphi_{j=0}(r,\theta) = C\left[\begin{array}{c} e^{-i\frac{\theta}{2}} \\ i\,\mathrm{sgn}(B)e^{+i\frac{\theta}{2}} \end{array}\right]\left(\frac{K_{\frac{1}{2}}(\lambda_1 r)}{K_{\frac{1}{2}}(\lambda_1 R)} - \frac{K_{\frac{1}{2}}(\lambda_2 r)}{K_{\frac{1}{2}}(\lambda_2 R)}\right) \tag{8.35}$$

with $E = 0$. The modified Bessel function $K_{1/2}(x) = \sqrt{\frac{\pi}{2x}}\mathrm{e}^{-x}$ and C is a normalized constant. Thus, when $\nu = \frac{1}{2}$ or $-\frac{1}{2}$, there always exists a stable solution of $j = 0$ with the energy eigenvalue exactly zero for an arbitrary R. Since the energy eigenvalue is independent of the radius R, the half quantum flux here is also called topological defect. The existence of the zero-energy mode is valid even for an irregular hole which can be deformed continuously into a point-like defect.

8.4.2 Wormhole Effect

This solution can be generalized to three dimensions. Consider a topological insulator with a cylindrical hole (say along z-direction) of radius R threaded by a magnetic flux $\nu = \frac{1}{2}$. We take the periodic boundary condition along the z-direction. Thus, k_z is a good quantum number. The three-dimensional effective Hamiltonian can be separated into two parts:

$$H_{3D}(k_z) = H_{2D} + V(k_z),\qquad (8.36)$$

where

$$H_{2D}(x, y) = \nu p_x \alpha_x + \nu p_y \alpha_y + \left[m\nu^2 - B(p_x^2 + p_y^2) \right]\beta,\qquad (8.37)$$

$$V(k_z) = \nu \hbar k_z \alpha_z - B k_z^2 \beta.\qquad (8.38)$$

At $k_z = 0$, $V(k_z = 0) = 0$. In this case, H_{3D} are equivalent to two separated two-dimensional Dirac equations in a hole threading a magnetic flux. Using the solutions in the last paragraph for two-dimensional, one obtains two solutions of zero energy:

$$\varphi_1 = C \begin{pmatrix} e^{-i\frac{\theta}{2}} \\ 0 \\ 0 \\ i\,\mathrm{sgn}(B)e^{+i\frac{\theta}{2}} \end{pmatrix} \left(\frac{K_{\frac{1}{2}}(\lambda_1 r)}{K_{\frac{1}{2}}(\lambda_1 R)} - \frac{K_{\frac{1}{2}}(\lambda_2 r)}{K_{\frac{1}{2}}(\lambda_2 R)} \right)\qquad (8.39)$$

and

$$\varphi_2 = C \begin{pmatrix} 0 \\ e^{+i\frac{\theta}{2}} \\ i\,\mathrm{sgn}(B)e^{-i\frac{\theta}{2}} \\ 0 \end{pmatrix} \left(\frac{K_{\frac{1}{2}}(\lambda_1 r)}{K_{\frac{1}{2}}(\lambda_1 R)} - \frac{K_{\frac{1}{2}}(\lambda_2 r)}{K_{\frac{1}{2}}(\lambda_2 R)} \right).\qquad (8.40)$$

The order of the base has been reorganized. Note the fact that the two separated equations are counter-partners of time reversal, and the prefactors of θ in φ_2 change a sign. Using these two solutions as the basis, one obtains an effective Hamiltonian for a nonzero k_z:

$$H_{\mathrm{eff}} = \mathrm{sgn}(B)\nu\hbar k_z \sigma_y.\qquad (8.41)$$

Thus, there exists a pair of gapless helical electron states along the hole or magnetic flux, which is independent of the radius R. This is so-called "wormhole" effect [12].

Dislocations are line defects of the three-dimensional crystalline order, characterized by a lattice vector $\mathbf{B}$ (the Burgers vector). This is rather like the quantized vorticity of a superfluid vortex and must remain constant over its entire length. Dislocations in three-dimensional crystal of topological insulator is equivalent to a hole threading a magnetic flux $\nu = \frac{1}{2}$. Ran et al. found that each dislocation induces a pair of one-dimensional modes bound to it, which propagate in opposite directions and traverse the bulk band gap [13].

8.4.3   Witten Effect

The Witten effect is a fundamental property of the axion media [14]. The idea of the axion was first introduced as a means to solve what is known as the strong charge-parity problem in the physics of strong interaction. After the discovery of topological insulator, Qi, Hughes and Zhang proposed that the electromagnetic response in topological insulator is characterized by an axion term, $\Delta L_{\text{axion}} = \theta \frac{e^2}{2\pi h} \mathbf{B} \cdot \mathbf{E}$ with $\theta = \pi$ [15]. The Witten effect means that a unit magnetic monopole $\phi_0 = h/e$ placed in a topological insulator will bind a fractional charge $Q = -e\left(n + \frac{1}{2}\right)$ with n integer. This effect has been already used to identify whether a system is topologically trivial or nontrivial by means of numerical calculation [16].

The axion term revises the Gauss' law and Ampere's law by adding extra source terms

$$\nabla \cdot \mathbf{D} = \rho - \frac{\alpha}{\pi \mu_0 c} \nabla \theta \cdot \mathbf{B}, \tag{8.42a}$$

$$\nabla \times \mathbf{H} = \partial_t \mathbf{D} + j + \frac{\alpha}{\pi \mu_0 c} \left(\nabla \theta \times \mathbf{E} + \partial_t \theta \mathbf{B} \right), \tag{8.42b}$$

where $\mathbf{D} = \epsilon_0 \mathbf{E} + \mathbf{P}$ and $\mathbf{H} = \frac{1}{\mu_0} \mathbf{B} - \mathbf{M}$. The fine structure constant $\alpha = e^2/(2\epsilon_0 h c)$. Suppose there is a point-like magnetic monopole situated at the origin of the strength ϕ_0. The static magnetic field is given by $\mathbf{B} = \frac{\phi_0}{r^2} \mathbf{r}$ or $\nabla \cdot \mathbf{B} = \phi_0 \delta(\mathbf{r})$. Suppose $\theta = 0$ initially and then increases to $\theta = \pi$. θ is uniform in space, and there is no current in vacuum. We take the divergence of revised Ampere's law:

$$\nabla \cdot \partial_t \mathbf{E} = -\frac{\alpha c}{\pi} \partial_t \theta \nabla \cdot \mathbf{B}. \tag{8.43}$$

Thus, when θ increases from 0 to π, integrating the equation yields

$$\nabla \cdot [\mathbf{E}(\theta = \pi) - \mathbf{E}(\theta = 0)] = -\frac{1}{\epsilon_0} \frac{e^2}{2h} \nabla \cdot \mathbf{B} = -\frac{1}{\epsilon_0} \frac{e}{2} \delta(\mathbf{r}). \tag{8.44}$$

This demonstrates that a magnetic monopole ϕ_0 can bind an extra fractional charge $-e/2$.

To understand the Witten effect, we consider a sphere with radius R of an isotropic topological insulator with a magnetic monopole $2q\phi_0$ situated at the origin:

$$H = v(\mathbf{p} + e\mathbf{A}) \cdot \alpha + \left[mv^2 - B(\mathbf{p} + e\mathbf{A})^2\right] \beta.$$

$$= \begin{pmatrix} mv^2 - B\Pi^2 & v\Pi \cdot \sigma \\ v\Pi \cdot \sigma & -mv^2 + B\Pi^2 \end{pmatrix}, \tag{8.45}$$

where $\Pi = \mathbf{p} + e\mathbf{A}$ and $\nabla \times \mathbf{A} = \frac{2q\phi_0}{r^2}\mathbf{r}$. It is well known that the magnetic field of a magnetic monopole cannot be derived from a single expression of vector potential valid everywhere. We can construct a pair of the vector potentials

$$\mathbf{A}^I = +\frac{2q\phi_0}{r}\frac{1 - \cos\theta}{\sin\theta}\hat{\phi}, \text{ for } \theta < \pi - \varepsilon, \tag{8.46a}$$

$$\mathbf{A}^{II} = -\frac{2q\phi_0}{r}\frac{1 + \cos\theta}{\sin\theta}\hat{\phi}, \text{ for } \theta > \varepsilon, \tag{8.46b}$$

such that there are no singularity in the two potentials in the defined range. In the overlapping region $\varepsilon < \theta < \pi - \varepsilon$, the two potentials are related by a gauge transformation:

$$\mathbf{A}^I - \mathbf{A}^{II} = \frac{4q\phi_0}{r\sin\theta}\hat{\phi}. \tag{8.47}$$

In the overlapping region, we can use either $\mathbf{A}^I$ or $\mathbf{A}^{II}$, the corresponding wave functions are related by a phase factor $\exp[i4q\pi]$. Thus, the single value condition for the wave function for either $\mathbf{A}^I$ or $\mathbf{A}^{II}$ implies $2q = $ integer, which is the quantization condition for a magnetic charge [17].

Following Kazama et al. [18], we can define

$$\mathbf{L} = \mathbf{r} \times \Pi - q\hbar\frac{\mathbf{r}}{r} \tag{8.48}$$

which satisfies the commutation relation of the orbital angular momentum, $[L_\alpha, L_\beta] = i\hbar\epsilon_{\alpha\beta\gamma}L_\gamma$. Denote Y_{q,l,l_z} as the eigenfunction of L^2 and L_z with the eigenvalues $l(l + 1)\hbar^2$ and $l_z\hbar$ ($l_z = -l, -l_z + 1, \cdots,$ and l). The total angular momentum $\mathbf{J}$ is defined as $\mathbf{J} = \mathbf{L} + \mathbf{S}$ where the spin $\mathbf{S} = \frac{1}{2}\hbar\sigma$. The eigenstates of $\mathbf{J}^2$ and $\mathbf{J}_z$ can be constructed by adding $\mathbf{L}$ and $\mathbf{S}$:

$$\phi_{j,j_z}^{(1)} = \begin{pmatrix} \sqrt{\dfrac{j + m}{2j}}Y_{q,l=j-1/2,j_z-1/2} \\ \\ \sqrt{\dfrac{j - m}{2j}}Y_{q,l=j-1/2,j_z+1/2} \end{pmatrix} \tag{8.49a}$$

$$
\phi_{j,j_z}^{(2)} = \begin{pmatrix} -\sqrt{\dfrac{j-m+1}{2j+2}} Y_{q,l=j+1/2,j_z-1/2} \\[4mm] \sqrt{\dfrac{j+m+1}{2j+2}} Y_{q,l=j+1/2,j_z+1/2} \end{pmatrix} \tag{8.49b}
$$

which are for $j = l + 1/2$ and $j = l - 1/2$, respectively. The coefficients in the expressions are the Clebsch-Gordan coefficients. For simplicity we here focus on the zero-energy solution. We construct an ansatz for the trivial wave function for $j = |q| - \frac{1}{2}$ and $l = j + \frac{1}{2} = |q|$,

$$
\Psi = \begin{pmatrix} F(r)\phi_{j,j_z}^{(2)} \\ G(r)\phi_{j,j_z}^{(2)} \end{pmatrix}. \tag{8.50}
$$

Substituting the trial wave function into the stationary equation of H in Eq. (8.45), and using the relation

$$
\sigma \cdot \Pi \phi_{j,j_z}^{(2)}(\theta,\phi) = -i\,\mathrm{sgn}(q)\hbar(\partial_r + r^{-1})\phi_{j,j_z}^{(2)}(\theta,\phi), \tag{8.51}
$$

the equation for the radial part of the wave function is reduced to

$$
\left[-i\,\mathrm{sgn}(q)v\hbar\partial_r\sigma_x + \left[mv^2 + B\hbar^2(\partial_r^2 - \frac{|q|}{r^2})\right]\sigma_z \right] \begin{pmatrix} rF(r) \\ rG(r) \end{pmatrix} = E \begin{pmatrix} rF(r) \\ rG(r) \end{pmatrix}. \tag{8.52}
$$

For our purpose, we consider a sphere of a large radius R enough by ignoring the finite size effect between the surface states and the bound states near the center.

When $r >> 1$, $\frac{|q|}{r^2} \to 0$. In this case, Eq. (8.52) is approximately reduced to the one-dimensional Dirac equation,

$$
\left[-i\,\mathrm{sgn}(q)v\hbar\partial_r\sigma_x + \left(mv^2 + B\hbar^2\partial_r^2 \right)\sigma_z \right] \begin{pmatrix} rF(r) \\ rG(r) \end{pmatrix} = E \begin{pmatrix} rF(r) \\ rG(r) \end{pmatrix} \tag{8.53}
$$

as in Eqs. (2.33) and (8.34), in which there always exists an end state solution of zero energy near $r = R$ when $mB > 0$. The solution has the form

$$
\begin{pmatrix} F(r) \\ G(r) \end{pmatrix} = \frac{C}{r} \left(\frac{e^{\lambda_1 r}}{e^{\lambda_1 R}} - \frac{e^{\lambda_2 r}}{e^{\lambda_2 R}} \right) \begin{pmatrix} 1 \\ i\eta \end{pmatrix} \tag{8.54}
$$

with $\eta = -\mathrm{sgn}(qBv)$, and $\lambda_{1,2} = \left|\frac{v}{2B\hbar}\right| \pm \sqrt{\frac{v^2}{4B^2\hbar^2} - \frac{mv^2}{B\hbar^2}}$. These solutions are valid even for complex $\lambda_{1,2}$.

Near the center of the sphere $r = 0$, we can find another solution:

$$\begin{pmatrix} F(r) \\ G(r) \end{pmatrix} = C' \frac{e^{-\zeta\rho/2}}{\sqrt{\rho}} J_{\sqrt{|q|+1/4}} \left(\sqrt{1 - \zeta^2/4}\rho \right) \begin{pmatrix} 1 \\ -i\eta \end{pmatrix}, \qquad (8.55)$$

where $\rho = \sqrt{m^*v^2/B\hbar^2}r$, $\zeta = 1/\sqrt{mB}$, and $J_\alpha(x)$ is the first Bessel function. C and C' are the normalization constants. From the asymptotic behavior of the first Bessel function, $J_\alpha(x) \to x^{|\alpha|}$, it concludes that the solution is convergent at $\rho \to 0$ when $q \neq 0$. For $\zeta^2 > 4$, $J_\alpha(x)$ is replaced by the modified Bessel function $K_{\sqrt{|q|+1/4}} \left(\sqrt{\zeta^2/4 - 1}\rho \right)$.

Since the final result is independent of the eigenvalue j_z, there are $2|q|$ ($= 2j + 1$)-fold degeneracy of the zero-energy states as well as the double degeneracy of the states near the center and the surface. For each j_z, the double degeneracy of the bound states can be lifted when the radius R is finite, and the two states at the center and the surface will be coupled to form two new states, in which one has a positive energy and the other has a negative energy. The energy gap decays exponentially in the radius R. In this case, each bound state is split into two halves: one half is distributed around the surface of the system, while another one surrounds the magnetic monopole. For a topological insulator, the system is half filled, and only $2|q|$ zero-energy states are occupied, while all other negative energy states are filled.

However, the double degeneracy of the zero-energy bound states for $q = 1/2$ and a large R makes it possible that the bound state near the center is either fully or partially occupied. The charge binding around the center is not determined. Therefore, it deserves further studying whether the electromagnetic response in topological insulator is really characterized by an axion term or not.

8.5 Disorder Effect to Transport

We come to discuss the effect of the in-gap bound states to the transport in topological insulators. The wave function of the in-gap bound state is localized around the vacancy or defect. Away from the center, the wave function decays exponentially, that is, $\propto e^{-r/\xi}$. The characteristic length ξ reflects the spatial distribution of the wave function. When two vacancies are close within the distance comparable with the characteristic length ξ, the overlapping of the wave functions in space becomes possible. Consequently electrons in one bound state has possibility to jump to another bound states.

For a single vacancy close to the boundary of the quantum spin Hall system, the edge states will be scattered by the in-gap bound state of the vacancy. However, if there is no other defects or disorders in the bulk, the electrons in the edge state will not be further scattered away from the edge as what happens in the quantum Hall effect [19], which also indicates the robustness of the edge states against the defects or disorders. The situation will change if the concentration of vacancies is dense

Fig. 8.5 Schematic of melting of quantum spin Hall effect due to the holes or defects. The helical edge states at different boundaries can be scattered via the in-gap bound states induced by the holes or defects

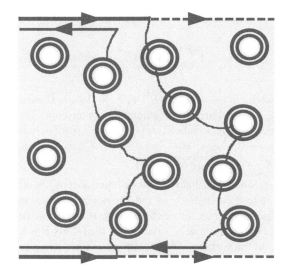

enough. The bound states could form an "impurity band" in the gap of bulk bands. When the wave functions of the bound states overlap in space as shown in Fig. 8.5, it becomes possible that the electrons in the edge state on one side can be scattered to the other side via a multiple scattering procedures. In this case, the backscattering of electrons in two sides occurs, and the quantum transport of the edge states will break down. Thus, there exists a critical point where the quantum percolation occurs due to the in-gap bound states of vacancies.

This picture can be demonstrated explicitly by calculating the conductance in a two-terminal setup of the quantum spin Hall system as a function of a concentration of vacancies. We use the open boundary condition with two edges and the periodic boundary condition or a cylinder without the edge states. In an open boundary condition, the calculated conductance is quantized to be $2e^2/h$. While it is immune to the low density of vacancies, the conductance decreases with the density of vacancies quickly, and the quantum spin Hall effect is destroyed completely (see Fig. 8.6a). In a cylinder or periodic boundary condition, the conductance is zero in a pure quantum spin Hall state as there is no edge state in the geometry. A nonzero conductance appears and increases with the concentration of vacancies, and reaches at a maximal for a specific value of the concentration. Then it decreases with increasing the concentration of vacancies (see Fig. 8.6b). Figure 8.6c–e shows the density of states at different concentrations. A nonzero peak appears at $E = 0$ near the critical concentration, which demonstrates the occurrence of quantum percolation and appearance of metallic phase. From the calculated conductance, it reveals a quantum phase transition from a quantum spin Hall state (Z_2: $\nu = 1$) to a conventional insulator (Z_2: $\nu = 0$).

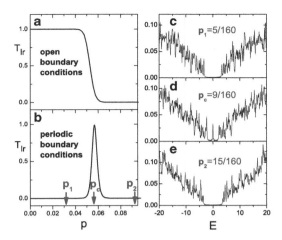

Fig. 8.6 Quantum percolation of electrons through in-gap bound states induced by randomly distributed vacancies or holes of size 1×1. The sample is $W \times L = 160 \times 160$ in size. (**a**) and (**b**) show the transmission coefficients T_{tr} vs. the concentration of vacancies p under the open boundary condition and periodic or closed boundary condition. (**c**)–(**e**) Are the density of states at the concentration $p_1 = 5/160$, $p_c = 9/160$, and $p_2 = 15/160$, respectively (Adapted from [20])

8.6 Further Reading

- Y. Ran, Y. Zhang, A. Vishwanath, One-dimensional topologically protected modes in topological insulators with lattice dislocations. Nat. Phys. **5**, 298 (2009)
- W.Y. Shan, J. Lu, H.Z. Lu, S.Q. Shen, Vacancy-induced in-gap bound states in topological insulators. Phys. Rev. B **84**, 035307 (2011)

References

1. L. Yu, Acta. Phys. Sin. **21**, 75 (1965)
2. H. Shiba, Prog. Theor. Phys. **40**, 435 (1968)
3. A.V. Balatsky, I. Vekhter, J.X. Zhu, Rev. Mod. Phys. **78**, 373 (2006)
4. J. Lu, W.Y. Shan, H.Z. Lu, S.Q. Shen, New J. Phys. **13**, 103016 (2011)
5. J.R. Schrieffer, *Theory of Superconductivity* (Persues Books, Reading, 1964)
6. W.Y. Shan, J. Lu, H.Z. Lu, S.Q. Shen, Phys. Rev. B **84**, 035307 (2011)
7. B.A. Bernevig, T.L. Hughes, S.C. Zhang, Science **314**, 1757 (2006)
8. H.Z. Lu, W.Y. Shan, W. Yao, Q. Niu, S.Q. Shen, Phys. Rev. B **81**, 115407 (2010)
9. W.Y. Shan, H.Z. Lu, S.Q. Shen, New J. Phys. **12**, 043048 (2010)
10. B. Zhou, H.Z. Lu, R.L. Chu, S.Q. Shen, Q. Niu, Phys. Rev. Lett. **101**, 246807 (2008)
11. Y. Zhang, K. He, C.Z. Chang, C.L. Song, L.L. Wang, X. Chen, J.F. Jia, Z. Fang, X. Dai, W.Y. Shan, S.Q. Shen, Q. Niu, X.L. Qi, S.C. Zhang, X.C. Ma, Q.K. Xue, Nat. Phys. **6**, 584 (2010)
12. G. Rosenberg, H.M. Guo, M. Franz, Phys. Rev. B **82**, 041104 (2010)
13. Y. Ran, Y. Zhang, A. Vishwanath, Nat. Phys. **5**, 298 (2009)
14. E. Witten, Phys. Lett. B **86**, 283 (1979)

15. X.L. Qi, T.L. Hughes, S.C. Zhang, Phys. Rev. B **78**, 195424 (2008)
16. G. Rosenberg, M. Franz, Phys. Rev. **82**, 035105 (2010)
17. T.T. Wu, C.N. Yang, Nucl. Phys. B **107**, 365 (1976)
18. Y. Kazama, C.N. Yang, A.S. Goldhaber, Phys. Rev. D **15**, 2287 (1977)
19. M. Büttiker, Phys. Rev. B **38**, 9375 (1988)
20. R.L. Chu, J. Lu, S.Q. Shen, EPL **100**, 17013 (2012)

Chapter 9
Topological Superconductors and Superfluids

Abstract Superfluid phases in liquid ^{3}He are the topological ones, which have the edge or surface states just like topological insulators. Spin-triplet superconductors are potential candidates of topological superconductors.

Keywords Helium three superfluid • p-wave pairing superconductor • Spin-triplet superconductor • Edge states • Surface states

The study of topological phases in superconductors and superfluids had a long history even before the birth of topological insulators. ^{3}He-B and ^{3}He-A phases are topological superfluid liquids and can be characterized by topological invariants [1]. A complex $p + ip$ wave pairing superconductor is also known to possess two topologically distinct phases [2]. Soon after the discovery of topological insulator, it was realized that there is an explicit analogy between topological insulator and superconductor because the particle-hole symmetry in the Bogoliubov-de Gennes (BdG) Hamiltonian for quasiparticles in superconductors is analogous to the time reversal symmetry in the Hamiltonian for a band insulator. The BdG Hamiltonians for a $p+ip$ wave superconductor and ^{3}He-B superfluids are identical to the modified Dirac equation as we discussed for topological insulators, although the bases of these equations are completely different.

Superconductivity is a quantum phenomenon that the resistivity in certain materials disappears below a characteristic temperature, which was discovered by H.K. Onnes in 1911 in Leiden [3]. A superconductor is characterized by zero resistance, Meissner effect or perfect diamagnetization, and magnetic flux quantization, though some physical properties vary from material to material, such as the heat capacity, the transition temperatures, and the critical fields. The existence of the universal properties in superconductors implies that superconductivity is a quantum phase having distinguishing properties which are largely independent of microscopic details. The theory of superconductivity was formulated by Bardeen, Cooper, and

S.-Q. Shen, *Topological Insulators: Dirac Equation in Condensed Matters*,
Springer Series in Solid-State Sciences 174, DOI 10.1007/978-3-642-32858-9_9,
© Springer-Verlag Berlin Heidelberg 2012

Schrieffer in 1957, and is called the BCS theory [4]. This theory has successfully described a large class of superconducting materials, such as aluminum.

The basic idea of the BCS theory is that electrons in the metal form bound pairs. Cooper pointed out that the ground state of a normal metal was unstable at zero temperature if the interaction between electrons near the Fermi surface is attractive. For an ideal metal, electrons at zero temperature form a Fermi sphere in the momentum space, which has a sharp step in energy. If there exists a weak attractive interaction between electrons near the Fermi surface, Cooper found that two electrons with opposite spins and momenta can forget the mutual scattering and form a bound state, which always has lower energy than that of two free electrons. In some metals, the electron-phonon interaction can provide such kind of attractive interaction near the Fermi surface. Most electrons in the Fermi sphere do not form the bound states, but only those within the Debye energy. The bound states of electrons pairs or Cooper pairs behave like bosons and can condensate at low temperatures, that is, Bose-Einstein condensation. The condensation of the Cooper pairs exhibits superconductivity, which requires a many-body description.

To explore topological phase in superconductor, we focus on the p-wave superconductivity.

9.1 Complex $(p + ip)$-Wave Superconductor of Spinless or Spin-Polarized Fermions

A complex p-wave spinless superconductor has two topologically distinct phases, one is the strong pairing phase and the other is the weak pairing phase [1, 2]. The weak pairing phase is identical to the Moore-Read quantum Hall state [2]. The system can be described by the modified Dirac model. In the BCS theory, the effective Hamiltonian for quasiparticles is

$$H_{\text{eff}} = \sum_{\mathbf{k}} \left[\xi_{\mathbf{k}} c_{\mathbf{k}}^{\dagger} c_{\mathbf{k}} + \frac{1}{2} \left(\Delta_{\mathbf{k}}^{*} c_{-\mathbf{k}} c_{\mathbf{k}} + \Delta_{\mathbf{k}} c_{\mathbf{k}}^{\dagger} c_{-\mathbf{k}}^{\dagger} \right) \right]. \tag{9.1}$$

It is noted that the electrons of $\mathbf{k}$ and $-\mathbf{k}$ are coupled together to form a Cooper pair. Though the number of electrons are not conserved in this effective Hamiltonian, the number parity, that is, the even or odd number of electrons, is conserved. For a small $\mathbf{k}$, we take $\xi_{\mathbf{k}} = \frac{k^2}{2m^*} - \mu$, where m^* is the effective mass and $-\mu$ is a constant of $\xi_{\mathbf{k}=0}$.

For the complex p-wave pairing, we take $\Delta_{\mathbf{k}}$ to be an eigenfunction of rotations in $\mathbf{k}$ with angular momentum l. For $l = +1$, at small $\mathbf{k}$ it generically takes the form

$$\Delta_{\mathbf{k}} = \Delta(\mathbf{k}_x + i\mathbf{k}_y). \tag{9.2}$$

For $l = -1$, $\Delta_k = \Delta(k_x - ik_y)$. The states of $\Delta_k = \Delta(k_x \pm ik_y)$ are degenerate. Consider the anticommutation relation of fermions, $c_k^\dagger c_k = 1 - c_k c_k^\dagger$. We take $\psi_k^\dagger = (c_k^\dagger, c_{-k})$, and then the effective Hamiltonian can be written in a compact form,

$$H_{\text{eff}} = \frac{1}{2}\sum_k \psi_k^\dagger h_{\text{eff}} \psi_k \tag{9.3}$$

by ignoring a constant. Here H_{eff} has the identical form of the Dirac equation

$$h_{\text{eff}} = \Delta\left(k_x\sigma_x \mp k_y\sigma_y\right) + \left(\frac{k^2}{2m} - \mu\right) \tag{9.4}$$

for $\Delta_k = \Delta(k_x \pm ik_y)$.

The normalized ground state has the form

$$|\Omega\rangle = \prod_k (u_k + v_k c_k^\dagger c_{-k}^\dagger)\,|0\rangle \tag{9.5}$$

where $|0\rangle$ is the vacuum state and the product runs over the distinct pairs of k and $-k$. The functions of u_k and v_k are complex and satisfy $|u_k|^2 + |v_k|^2 = 1$. We introduce the Bogoliubov transformation

$$\begin{pmatrix} \alpha_k \\ \alpha_{-k}^\dagger \end{pmatrix} = \begin{pmatrix} u_k & -v_k \\ -v_{-k}^* & u_{-k}^* \end{pmatrix} \begin{pmatrix} c_k \\ c_{-k}^\dagger \end{pmatrix} \tag{9.6}$$

where $\left\{\alpha_k, \alpha_{k'}^\dagger\right\} = \delta_{k,k'}$ and $\alpha_k\,|\Omega\rangle = 0$. The resulting Hamiltonian becomes

$$K_{\text{eff}} = \frac{1}{2}\sum_k \left(\alpha_k^\dagger, \alpha_{-k}\right) \begin{pmatrix} \varepsilon_k & 0 \\ 0 & -\varepsilon_k \end{pmatrix} \begin{pmatrix} \alpha_k \\ \alpha_{-k}^\dagger \end{pmatrix}$$

$$= \frac{1}{2}\sum_k \left(\varepsilon_k \alpha_k^\dagger \alpha_k - \varepsilon_k \alpha_{-k}\alpha_{-k}^\dagger\right). \tag{9.7}$$

with $\varepsilon_k = \sqrt{\xi_k^2 + |\Delta_k|^2} > 0$. The first term represents the particle excitation with a positive energy and the second term the hole excitations with a negative energy. Performing the particle-hole transformation, or making use of $\alpha_{-k}\alpha_{-k}^\dagger = 1 - \alpha_{-k}^\dagger\alpha_{-k}$, we have

$$K_{\text{eff}} = \sum_k \frac{1}{2}\varepsilon_k \left(\alpha_k^\dagger\alpha_k - 1 + \alpha_{-k}^\dagger\alpha_{-k}\right)$$

$$= \sum_k \varepsilon_k \alpha_k^\dagger\alpha_k - \sum_k \frac{1}{2}\varepsilon_k \tag{9.8}$$

as $\varepsilon_k = \varepsilon_{-k}$.

From the eigenstate equation,

$$[K_{\text{eff}}, \alpha_{\mathbf{k}}] = \varepsilon_{\mathbf{k}} \alpha_{\mathbf{k}}, \tag{9.9}$$

one obtains

$$\begin{pmatrix} \xi_{\mathbf{k}} & -\Delta_{\mathbf{k}}^* \\ -\Delta_{\mathbf{k}} & -\xi_{\mathbf{k}} \end{pmatrix} \begin{pmatrix} u_{\mathbf{k}} \\ v_{\mathbf{k}} \end{pmatrix} = \varepsilon_{\mathbf{k}} \begin{pmatrix} u_{\mathbf{k}} \\ v_{\mathbf{k}} \end{pmatrix}. \tag{9.10}$$

The solutions are

$$u_{\mathbf{k}} = \sqrt{\frac{1}{2}\left(1 + \frac{\xi_{\mathbf{k}}}{\varepsilon_{\mathbf{k}}}\right)}; \tag{9.11a}$$

$$v_{\mathbf{k}} = -\frac{\Delta_{\mathbf{k}}}{|\Delta_{\mathbf{k}}|}\sqrt{\frac{1}{2}\left(1 - \frac{\xi_{\mathbf{k}}}{\varepsilon_{\mathbf{k}}}\right)}. \tag{9.11b}$$

Here we choose a gauge that $u_{\mathbf{k}}$ is real and positive.

The Bogoliubov-de Gennes equation for $u_{\mathbf{k}}$ and $v_{\mathbf{k}}$ becomes

$$i\hbar \frac{\partial}{\partial t} \begin{pmatrix} u_{\mathbf{k}} \\ v_{\mathbf{k}} \end{pmatrix} = K_{\text{eff}} \begin{pmatrix} u_{\mathbf{k}} \\ v_{\mathbf{k}} \end{pmatrix} \tag{9.12a}$$

where

$$K_{\text{eff}} = \begin{pmatrix} \xi_{\mathbf{k}} & -\Delta_{\mathbf{k}}^* \\ -\Delta_{\mathbf{k}} & -\xi_{\mathbf{k}} \end{pmatrix} = -\Delta(k_x \sigma_x \pm k_y \sigma_y) + \xi_{\mathbf{k}} \sigma_z. \tag{9.13}$$

In this way, the Bogoliubov-de Gennes equation has the exact form of two-dimensional modified Dirac equation

$$K_{\text{eff}} = -\Delta \left(k_x \sigma_x \pm k_y \sigma_y\right) + \left(\frac{k^2}{2m} - \mu\right)\sigma_z. \tag{9.14}$$

It is noted that this effective Hamiltonian is distinct from that for electrons.

If we treat the Bogoliubov-de Gennes equation as one Hamiltonian as that for band insulator, we can introduce the topological invariant for K_{eff},

$$n_c = \pm \left[\text{sgn}(\mu) + \text{sgn}\left(\frac{1}{m}\right)\right]/2. \tag{9.15}$$

Since we take the mass of the spinless particles m positive, we conclude that for a positive $\mu(> 0)$ the Chern number is $+1$ (or -1) and for a negative μ the Chern number is 0. For $\mu = 0$, the Chern number is equal to one half, which is similar to the case of $m \to +\infty$ and a finite μ. If the quadratic term in ξ_k is neglected, we see that the topological property will change completely.

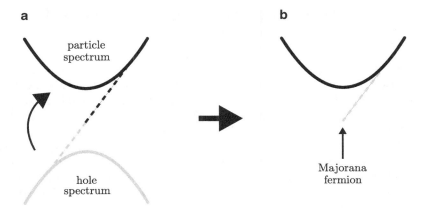

Fig. 9.1 (a) Particle-hole spectrum and edge-state spectrum of a non-trivial Bogoliubov-de Gennes equation for a weak pairing phase, or topologically nontrivial phase. (b) After the particle-hole transformation, the hole spectrum is merged into the particle spectrum. The zero energy mode is a Majorana fermion, $\gamma(E = 0) = \gamma^\dagger(E = 0)$

In general, from the solution of $u_\mathbf{k}$ and $v_\mathbf{k}$, we have three possibilities of behavior at small $\mathbf{k}$, $\varepsilon_\mathbf{k} - \xi_\mathbf{k} \to 0$.

1. $\xi_\mathbf{k} > 0$, in which $u_\mathbf{k} \to 1$ and $v_\mathbf{k} \to 0$. The BCS state is close to the vacuum, $|\Omega\rangle \to |0\rangle$.
2. $\xi_\mathbf{k} < 0$, in which $u_\mathbf{k} \to 0$ and $|v_\mathbf{k}| \to 1$. $|\Omega\rangle \to \prod_{\mathbf{k}'} v_\mathbf{k} c_\mathbf{k}^\dagger c_{-\mathbf{k}}^\dagger |0\rangle$ in which all the states with negative energy are occupied by free fermions.
3. $\xi_\mathbf{k} \to 0$, in which both $u_\mathbf{k}$ and $v_\mathbf{k}$ are nonzero.

Usually for a positive μ, the system is in the weak pairing phase, and for a negative μ, it is the strong pairing phase. Including the quadratic term in $\xi_\mathbf{k}$, we conclude that the weak pairing phase for positive μ is a typical topological insulator. Read and Green [2] argued that a bound-state solution exists at a straight domain wall parallel to the y-axis, with $\mu(r) = \mu(x)$ small and positive for $x > 0$, and negative for $x < 0$. There is only one solution for each k_y, and so we have a chiral Majorana fermions on the domain wall. From the two-dimensional solution, the system in a weak pairing phase should have a topologically protected and chiral edge state of Majorana fermions.

If we make the substitution in K_{eff}: $k_x \to -i\partial_x$ and $k_y \to -i\partial_y$, the BdG equation in Eq. (9.12a) has the identical form of the two-dimensional Dirac equation. When we find a solution for the edge state within the band gap, we emphasize the solution for $u_\mathbf{k}$, and $v_\mathbf{k}$ should satisfy the relation $\left|u_\mathbf{k}^2\right| + \left|v_\mathbf{k}^2\right| = 1$. For the vacuum, $u_\mathbf{k} = 1$ and $v_\mathbf{k} = 0$. The particle-hole spectra and the chiral edge spectra are presented in Fig. 9.1.

9.2   Spin-Triplet Pairing Superfluidity: ³He-A and ³He-B Phases

Helium has two isotopes, ³He and ⁴He. ⁴He atoms are bosons. At low temperatures, liquid ⁴He shows a phase transition to a superfluid state which is similar to the Bose-Einstein condensation, although strong inter-particle interaction should be taken into account. ³He atoms are fermions. Liquid ³He also shows a phase transition to a superfluid state, which is similar to the superconducting transition in a metal [5]. Since ³He atoms are neutral, there is no Meissner effect, but atoms form pairing like the Cooper pairs of electrons. Atoms also avoid the singlet pairing, as in metals, and tend to pair in the form of spin triplet, in which the spins align parallel [6]. A schematic of the phase diagram of ³He as a function of temperature and pressure is presented in Fig. 9.2.

9.2.1   ³He: Normal Liquid Phase

Before presenting the theory of superfluidity in ³He, we first briefly introduce a "normal" liquid phase of ³He atoms. The ³He atoms are charge neutral. Unlike the electrons in metals, these atoms are strongly interacting and highly correlated. According to the Fermi liquid theory, the low-lying excitations of the strongly interacting Fermi system can be described by a phenomenological model, in which the free energy of the system can be expanded in terms of low-energy excitation $\delta n_{\mathbf{p},\sigma}$

$$F = F_0 + \sum_{\mathbf{p},\sigma}(\epsilon_p - \mu)\delta n_{\mathbf{p},\sigma} + \frac{1}{2}\sum_{\mathbf{p},\sigma;\mathbf{p}',\sigma'} f_{\mathbf{p},\sigma;\mathbf{p}',\sigma'}\delta n_{\mathbf{p},\sigma}\delta n_{\mathbf{p}',\sigma'} + \cdots . \quad (9.16)$$

The parameters in this expression can be deduced from experiments such as specific heat, compressibility, sound velocity, and spin susceptibility. Here the energy zero

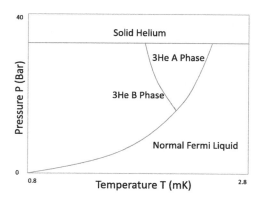

Fig. 9.2 Phase diagram of ³He in the low millikelvin temperature and pressure region

is defined as $-\mu$ at $p = 0$ such that $\epsilon_{p=0} = 0$ and $\epsilon_p = \frac{p^2}{2m^*}$ with the effective mass $m^* = 3m$, the three times of the bare mass of ^{3}He atom. The spin dependence of the effective interaction is written as

$$f_{\mathbf{p},\sigma;\mathbf{p}',\sigma'} = f^{(s)}_{\mathbf{p},\mathbf{p}'} + \sigma \cdot \sigma' f^{(t)}_{\mathbf{p},\mathbf{p}'}. \tag{9.17}$$

For details, the readers can refer to several excellent reviews of Fermi liquid theory such as Pine and Nozieres [7] and Leggett [8].

9.2.2 ^{3}He-B Phase

Theory of superconductivity for electrons in a spin-triplet state was developed by Balian and Werthamer [9], which succeeds in explaining superfluidity in ^{3}He. In their theory, fermions form spin-triplet pairs ($s = 1$), and the weak coupling between these pairs leads to condensate at low temperatures. The effective Hamiltonian for the quasiparticles has the form

$$H = \sum_{\mathbf{k},\sigma}(\epsilon_{\mathbf{k}} - \mu)c^{\dagger}_{\mathbf{k},\sigma}c_{\mathbf{k},\sigma} + \frac{1}{2}\sum_{\mathbf{k},\sigma;\mathbf{k}',\sigma',\mathbf{q}} V(\mathbf{q})c^{\dagger}_{\mathbf{k}+\mathbf{q},\sigma}c^{\dagger}_{\mathbf{k}'-\mathbf{q},\sigma'}c_{\mathbf{k}',\sigma'}c_{\mathbf{k},\sigma}. \tag{9.18}$$

The interaction potential describes the scattering process of the momentum change of two vectors $\mathbf{k}_i$ and $\mathbf{k}_f = \mathbf{k}_i + \mathbf{q}$. It can be expanded in spherical harmonics, and the first two terms are

$$V(\mathbf{q}) = V_0 + V_1\mathbf{k}_i \cdot \mathbf{k}_f + \cdots . \tag{9.19}$$

The first term is a repulsive s-wave interaction, $V_0 > 0$, and cannot cause the bound states. The second term is for a p-wave interaction, $V_1\mathbf{k}_i \cdot \mathbf{k}_f$, which leads to p-wave pairing. Thus, we only keep the second term for the theory of superfluidity. The interaction terms contains four operators. In the BCS theory, the atoms tend to form Cooper pairs, and the dominant interaction is reduced to the pair-pair interaction, $V(\mathbf{k}-\mathbf{k}')c^{\dagger}_{\mathbf{k},\sigma}c^{\dagger}_{-\mathbf{k},\sigma'}c_{-\mathbf{k}',\sigma'}c_{\mathbf{k}',\sigma}$. A mean-field approach is used to write the interaction term as a two-operator term by introducing the order parameters for pairing,

$$\begin{aligned}
H_{\text{eff}} = &\sum_{\mathbf{k},\sigma}(\epsilon_{\mathbf{k}} - \mu)c^{\dagger}_{\mathbf{k},\sigma}c_{\mathbf{k},\sigma} \\
&+ \frac{1}{2}\sum_{\mathbf{k}}\Delta_{+1}(\mathbf{k})c^{\dagger}_{\mathbf{k},\uparrow}c^{\dagger}_{-\mathbf{k},\uparrow} + \Delta^*_{+1}(\mathbf{k})c_{-\mathbf{k},\uparrow}c_{\mathbf{k},\uparrow} \\
&+ \sum_{\mathbf{k}}\Delta_0(\mathbf{k})c^{\dagger}_{\mathbf{k},\uparrow}c^{\dagger}_{-\mathbf{k},\downarrow} + \Delta^*_0(\mathbf{k})c_{-\mathbf{k},\downarrow}c_{\mathbf{k},\uparrow} \\
&+ \frac{1}{2}\sum_{\mathbf{k}}\Delta_{-1}(\mathbf{k})c^{\dagger}_{\mathbf{k},\downarrow}c^{\dagger}_{-\mathbf{k},\downarrow} + \Delta^*_{-1}(\mathbf{k})c_{-\mathbf{k},\downarrow}c_{\mathbf{k},\downarrow},
\end{aligned} \tag{9.20}$$

where three types of pairing order parameters are introduced,

$$\Delta_+(\mathbf{k}) = \sum_{\mathbf{k}'} V(\mathbf{k} - \mathbf{k}') \langle c_{-\mathbf{k}',\uparrow} c_{\mathbf{k}',\uparrow} \rangle \tag{9.21a}$$

$$\Delta_0(\mathbf{k}) = \sum_{\mathbf{k}'} V(\mathbf{k} - \mathbf{k}') \langle c_{-\mathbf{k}',\uparrow} c_{\mathbf{k}',\downarrow} \rangle \tag{9.21b}$$

$$\Delta_-(\mathbf{k}) = \sum_{\mathbf{k}'} V(\mathbf{k} - \mathbf{k}') \langle c_{-\mathbf{k}',\downarrow} c_{\mathbf{k}',\downarrow} \rangle \tag{9.21c}$$

and $\langle \cdots \rangle$ represents the thermodynamic average.

For a p-wave pairing, the order parameter $\Delta_m(\mathbf{k})$ is an odd function of momentum, $\Delta_m(-\mathbf{k}) = -\Delta_m(\mathbf{k})$. This condition can be derived from the definition

$$\Delta_+(-\mathbf{k}) = \sum_{\mathbf{k}'} V(-\mathbf{k} - \mathbf{k}') \langle c_{-\mathbf{k}',\uparrow} c_{\mathbf{k}',\uparrow} \rangle$$

$$= -\sum_{\mathbf{k}'} V(-\mathbf{k} - \mathbf{k}') \langle c_{\mathbf{k}',\uparrow} c_{-\mathbf{k}',\uparrow} \rangle = -\Delta_+(\mathbf{k}), \tag{9.22}$$

where the extra minus sign comes from the permutation of two operators in $\langle c_{-\mathbf{k}',\uparrow} c_{\mathbf{k}',\uparrow} \rangle$, when the interaction potential is even for a p-wave. Thus, order parameters $\Delta_m(\mathbf{k})$ have the p-wave symmetry and are proportional to the spherical harmonics, $Y_{1,-m}(\theta, \varphi)$,

$$\Delta_{+1}(\mathbf{k}) = \Delta(-k_x + i k_y); \tag{9.23a}$$

$$\Delta_0(\mathbf{k}) = \Delta k_z; \tag{9.23b}$$

$$\Delta_{-1}(\mathbf{k}) = \Delta(k_x + i k_y). \tag{9.23c}$$

In the lattice model, for example, on a cubic lattice, they are modified to fit the lattice symmetry,

$$\Delta_{+1}(\mathbf{k}) = \Delta(-\sin k_x + i \sin k_y); \tag{9.24a}$$

$$\Delta_0(\mathbf{k}) = \Delta \sin k_z; \tag{9.24b}$$

$$\Delta_{-1}(\mathbf{k}) = \Delta(\sin k_x + i \sin k_y). \tag{9.24c}$$

The pairing potential can be written in a compact form $V + V^\dagger$,

$$V = \sum_{\mathbf{k}} (c_{\mathbf{k},\uparrow}^\dagger, c_{\mathbf{k},\downarrow}^\dagger) \, \Xi(k) \begin{pmatrix} c_{-\mathbf{k},\downarrow}^\dagger \\ -c_{-\mathbf{k},\uparrow}^\dagger \end{pmatrix}, \tag{9.25a}$$

$$V^\dagger = \sum_{\mathbf{k}} (c_{-\mathbf{k},\downarrow}, -c_{\mathbf{k},\uparrow}) \, \Xi^\dagger(k) \begin{pmatrix} c_{\mathbf{k},\uparrow} \\ c_{\mathbf{k},\downarrow} \end{pmatrix}, \tag{9.25b}$$

where

$$\Xi(\mathbf{k}) = \begin{pmatrix} \Delta_0(\mathbf{k}) & -\Delta_{+1}(\mathbf{k}) \\ \Delta_{-1}(\mathbf{k}) & -\Delta_0(\mathbf{k}) \end{pmatrix}$$

$$= \begin{pmatrix} \Delta k_z & \Delta(k_x - ik_y) \\ \Delta(k_x + ik_y) & -\Delta k_z \end{pmatrix}$$

$$= \Delta(k_x \sigma_x + k_y \sigma_y + k_z \sigma_z). \tag{9.26}$$

We introduce the basis

$$\psi_{\mathbf{k}}^{\dagger} = (c_{\mathbf{k},\uparrow}^{\dagger}, c_{\mathbf{k},\downarrow}^{\dagger}, c_{-\mathbf{k},\downarrow}, -c_{-\mathbf{k},\uparrow}). \tag{9.27}$$

The effective Hamiltonian has the form

$$H = \frac{1}{2} \sum_{\mathbf{k}} \psi_{\mathbf{k}}^{\dagger} H_{\text{eff}} \psi_{\mathbf{k}} \tag{9.28}$$

where

$$H_{\text{eff}} = \Delta \left(k_x \alpha_x + k_y \alpha_y + k_z \alpha_z \right) + \xi_{\mathbf{k}} \beta \tag{9.29}$$

is identical to the modified Dirac equation.

Since this Hamiltonian is identical to the one for three-dimensional topological insulator, there exists a solution of the surface states near the boundary of the surface if it satisfies the condition for the topologically nontrivial phase. However, the bases of the fermion operators are quite different. In ^{3}He-B phase, we have particle and hole excitations, while we have the conduction bands and valence bands in topological insulator. Due to the particle-hole symmetry in the effective Hamiltonian, the particle and hole excitations always appear in pairs with energy, $\pm E$, which are connected by a particle-hole transformation

$$\gamma(E, \mathbf{k}) \rightarrow \gamma^{\dagger}(-E, -\mathbf{k}). \tag{9.30}$$

Therefore, in ^{3}He-B phase, the surface state consists of only one half Dirac cone with positive energy [10].

9.2.3 ^{3}He-A Phase: Equal Spin Pairing

When $\Delta_0(\mathbf{k}) = 0$, there exists a state with equal spin pairing. In this case, there is no relation between the orbital momentum of $\Delta_{+1}(\mathbf{k})$ and $\Delta_{-1}(\mathbf{k})$. Thus, the orbital motions of spin-up and spin-down particles are arbitrary. We can write them as

$$\Delta_{+1} = \Delta(k_x + ik_y); \tag{9.31a}$$

$$\Delta_{-1} = \Delta(k_x' + ik_y'). \tag{9.31b}$$

For the particles with spin up, the effective Hamiltonian is

$$H_\uparrow = \frac{1}{2} \sum_{\mathbf{k}} (c_{\mathbf{k},\uparrow}^\dagger, c_{-\mathbf{k},\uparrow}) \begin{pmatrix} \xi_{\mathbf{k}} & \Delta(k_x + ik_y) \\ \Delta(k_x - ik_y) & -\xi_{\mathbf{k}} \end{pmatrix} \begin{pmatrix} c_{\mathbf{k},\uparrow} \\ c_{-\mathbf{k},\uparrow}^\dagger \end{pmatrix}$$

$$= \frac{1}{2} \sum_{\mathbf{k}} (c_{\mathbf{k},\uparrow}^\dagger, c_{-\mathbf{k},\uparrow}) (\Delta k_x \sigma_x - \Delta k_y \sigma_y + \xi_{\mathbf{k}} \sigma_z) \begin{pmatrix} c_{\mathbf{k},\uparrow} \\ c_{-\mathbf{k},\uparrow}^\dagger \end{pmatrix} \tag{9.32}$$

where

$$\xi_k = \frac{\hbar^2}{2m}(k_x^2 + k_y^2) - \left(\mu - \frac{\hbar^2 k_z^2}{2m}\right). \tag{9.33}$$

This is identical to the two-dimensional modified Dirac equation. For a layered system, the term $\frac{\hbar^2 k_z^2}{2m}$ may be suppressed. The spectrum of the quasiparticle is

$$\varepsilon_{\mathbf{k}} = \sqrt{|\Delta|^2 (k_x^2 + k_y^2) + \xi_{\mathbf{k}}^2}. \tag{9.34}$$

The new feature of this model is that the effective chemical potential becomes k_z dependent, $\mu(k_z) = \mu - \frac{\hbar^2 k_z^2}{2m}$. The Chern number for a specific k_z is

$$n_c(k_z) = \begin{cases} 1 & \text{if } \dfrac{\hbar^2 k_z^2}{2m} < \mu \\[2mm] \dfrac{1}{2} & \text{if } \dfrac{\hbar^2 k_z^2}{2m} = \mu \\[2mm] 0 & \text{if } \dfrac{\hbar^2 k_z^2}{2m} > \mu \end{cases}. \tag{9.35}$$

At $k_z^2 = 2m\mu/\hbar^2$, $\varepsilon_k = |\Delta_1| k_\parallel + O(k_\parallel^2)$, which is linear in the momentum for a small $k_\parallel$ ($k_\parallel^2 = k_x^2 + k_y^2$). It is a marginal phase between two topologically distinguishing phases. Thus, in ^{3}He-A phase, there always exists a nodal point. Due to the nonzero Chern number, there exist chiral edge states around the boundary of system.

There are several possible choices in the state of equal spin pairing.

The Anderson-Brinkman-Morel state [11]:

$$\Delta_{+1} = \Delta_\alpha(\mathbf{k})(k_x + ik_y), \tag{9.36a}$$

$$\Delta_0 = 0, \tag{9.36b}$$

$$\Delta_{-1} = \Delta_\alpha(\mathbf{k})(k_x + ik_y), \tag{9.36c}$$

where $\Delta_\alpha(\mathbf{k})$ is an even function of k. In this case, the two phases of spin-up and spin-down particles are identical and possess the same Chern number if they are topologically nontrivial.

The two-dimensional planar state:

$$\Delta_{+1} = \Delta_\alpha(\mathbf{k})(k_x + ik_y), \tag{9.37a}$$

$$\Delta_0 = 0, \tag{9.37b}$$

$$\Delta_{-1} = \Delta_\alpha(\mathbf{k})(k_x - ik_y). \tag{9.37c}$$

In this case, the two phases of spin-up and spin-down particles possess opposite Chern numbers if they are topologically nontrivial.

The one-dimensional polar state:

$$\Delta_{+1} = 0, \tag{9.38a}$$

$$\Delta_0 = \Delta_\alpha(\mathbf{k})k_z, \tag{9.38b}$$

$$\Delta_{-1} = 0. \tag{9.38c}$$

The effective Hamiltonian becomes

$$H_{\text{eff}} = \Delta_\alpha(\mathbf{k})k_z\alpha_z + \xi(k_\parallel, k_z)\beta \tag{9.39}$$

where $\xi(k_\parallel, k_z) = \frac{\hbar^2}{2m}k_z^2 - (\mu - \frac{\hbar^2 k_\parallel^2}{2m})$. This equation can be deduced into two degenerate one-dimensional Dirac equation as discussed in Chap. 2. There always exist two crossing points at $\frac{\hbar^2 k_\parallel^2}{2m} = \mu$ and $k_z = 0$.

9.3 Spin-Triplet Superconductor: Sr_2RuO_4

There are several classes of candidates of spin-triplet superconductors, such as heavy fermion superconductor UPt_3, organic superconductor $(TMTSF)_2X$ ($X = ClO_4$ and PF_6), and ruthenate superconductors Sr_2RuO_4. In this section, we briefly introduce the unconventional properties of Sr_2RuO_4, which is considered most probably as a spin-triplet superconductor or even a topological superconductor, comparable with the odd-parity, pseudo-spin-triplet superconductor UPt_3.

Sr_2RuO_4 is an oxide superconductor that has the same layered structure as high-Tc cuprates but has a low superconducting transition temperature of 1.5 K [12]. The availability of high-quality single crystal and the relative simplicity of its fully characterized Fermi surface promoted a large number of experimental as well as theoretical studies. Rice and Sigrist [13] proposed the similarity between the superconductivity of Sr_2RuO_4 and the spin-triplet superfluidity of ^3He soon after the discovery of the ruthenate superconductivity, which leads to the first direct experimental evidence of spin-triplet pairing in Sr_2RuO_4 by the measurement of electron spin susceptibility with NMR.

At low temperatures, Sr_2RuO_4 maintains a tetragonal structure with the crystal point group symmetry D_{4h}. Neglecting the dispersion along the out-of-plane c direction, possible spin-triplet states are limited to those for the two-dimensional square lattice with C_{4v} symmetry. One possible state in Sr_2RuO_4 is so-called chiral pairing states, which possess two polarizations of relative orbital angular momentum of pairing quasiparticles: left and right polarizations correspond to

$$\Delta_0 \propto \sin k_x \pm i \sin k_y, \tag{9.40}$$

respectively. They are the states with the orbital angular momentum $L_z = +1$ and -1, and the Cooper pair spins lie in the plane, $S_z = 0$, while the total spin is $S = 1$.

The direct evidence of spin-triplet pairing in Sr_2RuO_4 is based on the electron spin susceptibility measurement by the NMR Knight shift of both ^{17}O and ^{99}Ru nuclei [14]. Combined with the observation of internal magnetic field by μSR, it is believed that the superconducting state of Sr_2RuO_4 is a spin-triplet chiral p-wave state, a two-dimensional analogue of the ^3He-A phase. The odd parity of the orbital part of the order parameter has been unambiguously demonstrated by phase sensitive measurements.

In the sector of $S = 1$ and $S_z = 0$, the superconducting state with $\Delta_0^\pm = \Delta(\sin k_x \pm i \sin k_y)$ is similar to spinless $p + ip$ wave superconductor. The Chern number can be defined as we discussed in Sect. 9.1. The states with $\Delta_0^+ = \Delta(\sin k_x + i \sin k_y)$ and $\Delta_0^- = \Delta(\sin k_x - i \sin k_y)$ are degenerate, but may have opposite Chern numbers due to the sign difference in $\Delta_0^\pm$. According to the bulk-edge correspondence, nonzero Chern number will lead to the emergence of the chiral edge states around the system boundary, which breaks time reversal symmetry. The superconducting states of Δ_0^+ and Δ_0^- possess opposite propagating edge states, respectively. The superconducting state in Sr_2RuO_4 has broken the time reversal symmetry spontaneously, and one of the states of $\Delta_0^\pm$ will be selected to be the ground state.

The existence of edge states has been studied in an experiment of quasi-particle tunneling spectroscopy [15]. The measured conductance spectra have revealed the evidence of the edge states. However, it is still under debate whether Sr_2RuO_4 is a topological superconductor or not. We expect more and conclusive experiments to settle down the issue in the near future [16].

9.4   Superconductivity in Doped Topological Insulators

Doped topological insulator $Cu_xBi_2Se_3$ exhibits the signature of superconductivity at low temperatures [17, 18]. The undoped Bi_2Se_3 compound is a topological insulator with a single Dirac cone of the surface states. Copper atoms can add holes or electrons in the Bi_2Se_3 lattice. It was found that about 10 % copper is needed to bring about the superconductivity in bulk Bi_2Se_3, in which the transition temperature of T_c is about 3.8 K, and was confirmed by the observation of the Meissner effect. The temperature dependence of specific heat suggests a fully gapped superconducting state. Experimental data even suggests the coexistence of superconductivity and the surface states protected by time reversal symmetry. However, superconductivity of doped topological insulator does not mean that the superconducting phase is always topologically nontrivial.

For a time reversal invariant superconductor, the mean-field Hamiltonian in Bogoliubov-de Gennes formalism preserves the additional particle-hole symmetry, $PH(\mathbf{k})P = -H(-\mathbf{k})$ with $P^2 = 1$. This particle-hole symmetry can define a Z_2 invariant as that for time reversal symmetry. Based on the calculation of the Z_2 invariant, Fu and Berg [19] and Sato [20] proposed that a time reversal-invariant centrosymmetric superconductor is a topological superconductor if (1) it has odd-parity pairing symmetry with a full superconducting gap and (2) its Fermi surface encloses an odd number of time reversal invariant momenta Γ_α (which satisfy $\Gamma_\alpha = -\Gamma_\alpha$ up to a reciprocal lattice vector) in the Brillouin zone.

It follows from the criteria that $Cu_x Bi_2Se_3$ is thought to be one of the potential candidates as a topological superconductor, which still needs more experiments to confirm.

9.5   Further Reading

- J.R. Schrieffer, *Theory of Superconductor* (Persues books, 1964)
- A.J. Leggett, Nobel lecture: superfluid ^{3}He: the early days as seen by a theorist. Rev. Mod. Phys. **76**, 909 (2004)
- N. Read, D. Green, Paired states of fermions in two dimensions with breaking of parity and time reversal symmetries and the fractional quantum Hall effect. Phys. Rev. B **61**, 10267 (2000)
- G.E. Volovik, *The Universe in a Helium Droplet* (Clarendon, Oxford, 2003)
- Y. Maeno, S. Kittaka, T. Nomura, S. Yonezawa, K. Ishida, Evaluation of spin-triplet superconductivity in Sr_2RuO_4. J. Phys. Soc. Jpn. **81**, 011009 (2012)

References

1. G.E. Volovik, *The Universe in a Helium Droplet* (Clarendon, Oxford, 2003)
2. N. Read, D. Green, Phys. Rev. B **61**, 10267 (2000)
3. H.K. Onnes, Commun. Phys. Lab. Univ. Leiden **12**, 120b (1911)
4. J. Bardeen, L.N. Cooper, J.R. Schrieffer, Phys. Rev. **106**, 162 (1957)
5. D.D. Osheroff, R.C. Richardson, D.M. Lee, Phys. Rev. Lett. **28**, 885 (1972)
6. J.C. Wheatley, Rev. Mod. Phys. **47**, 415 (1975)
7. D. Pine, P. Nozières, *The Theory of Quantum Liquids* (Benjamin, New York, 1966)

8. A.J. Leggett, Rev. Mod. Phys. **47**, 331 (1975)
9. R. Balian, N.R. Werthamer, Phys. Rev. **131**, 1553 (1963)
10. S.B. Chung, S.C. Zhang, Phys. Rev. Lett. **103**, 235301 (2009)
11. P.W. Anderson, P. Morel, Phys. Rev. **123**, 1911 (1961)
12. A.P. Mackenzie, Y. Maeno, Rev. Mod. Phys. **75**, 657 (2003)
13. T.M. Rice, M. Sigrist, J. Phys. Condens. Matter **7**, L643 (1995)
14. K. Ishida, M. Mukuda, Y. Kitaoka, K. Asayama, Z. Q. Mao, Y. Mori, Y. Maeno, Nature (London) **396**, 658 (1998).
15. S. Kashiwaya, H. Kashiwaya, H. Kambara, T. Furuta, H. Yaguchi, Y. Tanaka, Y. Maeno, Phys. Rev. Lett. **107**, 077003 (2011)
16. Y. Maeno, S. Kittaka, T. Nomura, S. Yonezawa, K. Ishida, J. Phys. Soc. Jpn. **81**, 011009 (2012)
17. Y.S. Hor, A.J. Williams, J.G. Checkelsky, P. Roushan, J. Seo, Q. Xu, H.W. Zandbergen, A. Yazdani, N.P. Ong, R.J. Cava, Phys. Rev. Lett. **104**, 057001 (2010)
18. L.A. Wray, S.-Y. Xu, Y. Xia, Y.S. Hor, D. Qian, A.V. Fedorov, H. Lin, A. Bansil, R.J. Cava, M.Z. Hasan, Nat. Phys. **6**, 855 (2010)
19. L. Fu, E. Berg, Phys. Rev. Lett. **105**, 097001 (2010)
20. M. Sato, Phys. Rev. B **81**, 220504(R) (2010)

Chapter 10
Majorana Fermions in Topological Insulators

Abstract A Majorana fermion is a particle that is its own antiparticle. This type of particles can appear as an end state in one-dimensional topological superconductor or the bound state induced by a half-quantized vortex in two-dimensional topological superconductors.

Keywords Majorana fermion • Kitaev model • Non-Abelian statistics • Quasi-one-dimensional p-wave superconductor

In his interpretation of the Dirac equation, Dirac introduced the concept of antiparticle for the negative energy solution. While the positive energy solution is used to describe an electron with spin $\frac{1}{2}$, the negative energy solution is for an antiparticle for electron, i.e., positron, which has a negative mass and positive elementary charge [1]. Ettore Majorana found that the Dirac equation can be separated into a pair of real wave equations, in which the fields are real and the particle and its antiparticle have no distinction [2]. For massless and neutral particles, their own antiparticle might be themselves. Neutrino and antineutrino are expected to be the same particles. However, Majorana fermions as elementary particles have not yet been realized in Nature [3]. Now it is highly possible to realize Majorana fermions in solids as quasiparticles of collective behaviors of many-particle systems.

10.1 What Is the Majorana Fermion?

A Majorana fermion satisfies the rules

$$\gamma_i^\dagger = \gamma_i \tag{10.1}$$

and

$$\{\gamma_i, \gamma_j^\dagger\} = \gamma_i \gamma_j^\dagger + \gamma_j^\dagger \gamma_i = \delta_{ij}. \tag{10.2}$$

A fermion operator can be always written in terms of two Majorana fermions,

$$c_{12}^\dagger = \frac{1}{\sqrt{2}}(\gamma_1 + i\gamma_2), \tag{10.3a}$$

$$c_{12} = \frac{1}{\sqrt{2}}(\gamma_1 - i\gamma_2), \tag{10.3b}$$

with $\gamma_1^\dagger = \gamma_1$ and $\gamma_2^\dagger = \gamma_2$. Conversely,

$$\gamma_1 = \frac{1}{\sqrt{2}}(c_{12}^\dagger + c_{12}), \tag{10.4a}$$

$$\gamma_2 = \frac{1}{i\sqrt{2}}(c_{12}^\dagger - c_{12}). \tag{10.4b}$$

One γ_i changes the fermion number between even and odd, which is called the fermion number parity. The fermion parity operator is

$$P = 1 - 2c_{12}^\dagger c_{12} = 2i\gamma_1\gamma_2, \tag{10.5}$$

which has an eigenvalue $+1$ if the state is empty, and -1 if the state is occupied.

10.2   Majorana Fermions in p-Wave Superconductors

10.2.1   Zero-Energy Mode Around a Quantum Vortex

The quantum flux in the p-wave superconductor can induce a bound state of zero energy, which is a Majorana fermion. Consider a hole of a radius R through which a magnetic flux ϕ threads. We require that the wave function vanishes at $r = R$. Due to the existence of the magnetic flux, the wave function should satisfy the boundary condition

$$\psi(\theta + 2\pi) = e^{i2\pi\phi/\phi_0}\psi(\theta) \tag{10.6}$$

where the quantum flux $\phi_0 = h/e$ if we take a gauge that the vector potential is absent in the Hamiltonian,

$$H(\mathbf{p} - e\mathbf{A}) \to H(\mathbf{p}). \tag{10.7}$$

In the polar coordinate system, the Hamiltonian becomes

$$H = \begin{pmatrix} -\dfrac{\hbar^2}{2m}(\partial_r^2 + \dfrac{1}{r}\partial_r + \dfrac{1}{r^2}\partial_\theta^2) - \mu & -i\,\Delta_0 e^{-i\theta}(\partial_r - \dfrac{i}{r}\partial_\theta) \\ -i\,\Delta_0 e^{+i\theta}(\partial_r + \dfrac{i}{r}\partial_\theta) & \dfrac{\hbar^2}{2m}(\partial_r^2 + \dfrac{1}{r}\partial_r + \dfrac{1}{r^2}\partial_\theta^2) + \mu \end{pmatrix}. \tag{10.8}$$

The wave function has the form

$$\psi = \begin{pmatrix} f(r)e^{i\nu\theta} \\ g(r)e^{i(\nu+1)\theta} \end{pmatrix}, \tag{10.9}$$

where $\nu = m + \phi/\phi_0$ and m is an integer. In this way, this two-dimensional problem is reduced to a one-dimensional equation for the radial part of the wave function

$$\begin{pmatrix} -\dfrac{\hbar^2}{2m}(\partial_r^2 + \dfrac{\partial_r}{r} - \dfrac{\nu^2}{r^2}) - \mu & -i\,\Delta_0(\partial_r + \dfrac{\nu+1}{r}) \\ -i\,\Delta_0(\partial_r - \dfrac{\nu}{r}) & \dfrac{\hbar^2}{2m}(\partial_r^2 + \dfrac{\partial_r}{r} - \dfrac{(\nu+1)^2}{r^2}) + \mu \end{pmatrix} \begin{pmatrix} f \\ g \end{pmatrix} = E \begin{pmatrix} f \\ g \end{pmatrix}. \tag{10.10}$$

The solution of the equation has the form

$$f = c_\nu^1 K_\nu(G_+ r) + c_\nu^2 K_\nu(G_- r), \tag{10.11a}$$

$$g = d_\nu^1 K_{\nu+1}(G_+ r) + d_\nu^2 K_{\nu+1}(G_- r), \tag{10.11b}$$

where $K_\nu(x)$ is the modified Bessel function of the second kind, and

$$G_\pm^2 = F \pm \sqrt{F^2 - \dfrac{4m^2}{\hbar^4}(\mu^2 - E^2)} \tag{10.12}$$

where $F = 2m^2\Delta_0^2/\hbar^4 - 2m\mu/\hbar^2$. With the boundary condition $\phi(r = R) = 0$, we have

$$\dfrac{G_+^2 + 2m(E+\mu)/\hbar^2}{G_+} \dfrac{K_{\nu+1}(G_+ R)}{K_\nu(G_+ R)} = \dfrac{G_-^2 + 2m(E+\mu)/\hbar^2}{G_-} \dfrac{K_{\nu+1}(G_- R)}{K_\nu(G_- R)}. \tag{10.13}$$

Solving the set of Eqs. (10.12) and (10.13), we may obtain the energy eigenvalues of the bound states around the hole. It is known that the equation becomes topologically nontrivial when $\mu > 0$. For $\mu < 0$, no bound state exists around the hole. For $\mu > 0$, there exist a series of the bound states. For a half quantum flux $\phi/\phi_0 = \frac{1}{2}$, there always exists a zero-energy mode which is independent of the radius of hole and robust against other interactions and even geometry of the hole (Fig. 10.1).

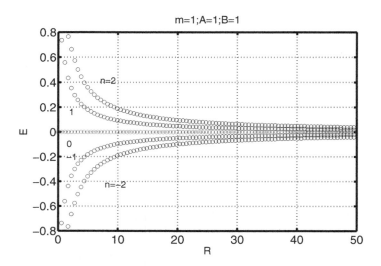

Fig. 10.1 The energy eigenvalues of the bound states as a function of radius R of the hole with $n = \nu + 1/2 = 0, \pm 1$, and ± 2. The parameters for numerical calculations: $\mu = \frac{1}{2m} = \Delta_0 = 1$ $(M = B = A = 1)$

The robustness of the zero mode for a half-quantum vortex can be demonstrated in the following way. When $\nu = -\frac{1}{2}$, Eq. (10.10) becomes

$$\left[-i \Delta_0 (\partial_r + \frac{1}{2r}) \sigma_x - \left[\frac{\hbar^2}{2m} \left(\partial_r^2 + \frac{\partial_r}{r} - \frac{1}{4r^2} \right) + \mu \right] \sigma_z \right] \begin{pmatrix} f \\ g \end{pmatrix} = E \begin{pmatrix} f \\ g \end{pmatrix}. \tag{10.14}$$

Furthermore, take a substitution,

$$\begin{pmatrix} f(r) \\ g(r) \end{pmatrix} = \frac{1}{\sqrt{r}} \varphi(r), \tag{10.15}$$

then the equation for the radial part of the wave function is reduced to

$$\left[-i \Delta_0 \partial_r \sigma_x - \left(\frac{\hbar^2}{2m} \partial_r^2 + \mu \right) \sigma_z \right] \varphi(r) = E \varphi(r) \tag{10.16}$$

which is identical to a one-dimensional modified Dirac equation. Thus, a zero mode may exist near $r = R$ if $\mu > 0$.

In a geometry of a disk of a finite radius, the solution of zero energy is split into two parts: one half is located around $r = 0$, the other half is distributed around the boundary. Thus, in this case Majorana fermion is non-local.

Ivanov [4] pointed out the equivalence between a half-quantum vortex for spinful fermions and a single-quantum vortex for spinless fermions, and there exists a zero-energy mode near a half-quantum vortex.

10.2.2 Majorana Fermions in Kitaev's Toy Model

The Kitaev's toy model is a one-dimensional chain of spinless p-wave superconductor [5],

$$H = -\mu \sum_{x=1}^{N} c_x^\dagger c_x - \sum_{x=1}^{N-1} (t c_x^\dagger c_{x+1} + \Delta e^{i\phi} c_x c_{x+1} + h.c.) \tag{10.17}$$

where $\mu, t > 0$, and $\Delta e^{i\phi}$ denote the chemical potential, the tunneling strength, and superconducting order parameters. Its BdG equation has the identical form of the modified Dirac model on a one-dimensional lattice. In the special case, $\mu = 0$ and $t = \Delta$, the Hamiltonian is reduced to

$$H = -t \sum_{x=1}^{N-1} (e^{i\phi/2} c_x + e^{-i\phi/2} c_x^\dagger)(e^{i\phi/2} c_{x+1} - e^{-i\phi/2} c_{x+1}^\dagger). \tag{10.18}$$

We define

$$\gamma_{B,x} = \frac{1}{\sqrt{2}}(e^{i\phi/2} c_x + e^{-i\phi/2} c_x^\dagger), \tag{10.19a}$$

$$\gamma_{A,x} = \frac{1}{i\sqrt{2}}(e^{i\phi/2} c_x - e^{-i\phi/2} c_x^\dagger) \tag{10.19b}$$

which are Majorana fermions and obey

$$\gamma_{A,x} = \gamma_{A,x}^\dagger, \tag{10.20a}$$

$$\gamma_{B,x} = \gamma_{B,x}^\dagger. \tag{10.20b}$$

In this way,

$$H = -2it \sum_{x=1}^{N-1} \gamma_{B,x} \gamma_{A,x+1}. \tag{10.21}$$

Two Majorana fermions $\gamma_{B,x}$ and $\gamma_{A,x+1}$ can combine to form a new fermion operator $d_x = \frac{1}{\sqrt{2}}(\gamma_{A,x+1} + i\gamma_{B,x})$, and $i\gamma_{B,x}\gamma_{A,x+1} = \frac{1}{2} - d_x^\dagger d_x$. The Hamiltonian becomes

Fig. 10.2 Schematic of two end Majorana states in the Kitaev's toy model

$$H = +2t \sum_{x=1}^{N-1} d_x^\dagger d_x - (N-1)t. \tag{10.22}$$

However, while all pairs of $(\gamma_{B,x}, \gamma_{A,x+1})$ for $x = 1, 2, \ldots, N-1$ form new fermions, $\gamma_{A,1}$ and $\gamma_{B,N}$ are absent from the Hamiltonian, i.e., $[\gamma_{A,1}, H] = [\gamma_{B,N}, H] = 0$. For $t > 0$, the lowest energy state is $|g\rangle$, in which $d_x |g\rangle = 0$ for all x, and

$$H |g\rangle = -(N-1)t |g\rangle . \tag{10.23}$$

Since $[\gamma_{A,1}, H] = [\gamma_{B,N}, H] = 0$, we can construct two degenerate states $\gamma_{A,1} |g\rangle$ and $\gamma_{B,N} |g\rangle$, which are related to an ordinary zero-energy fermion $d = \frac{1}{\sqrt{2}} (\gamma_{A,1} + i \gamma_{B,N})$. Since the γ operator changes the fermion parity one has $\langle g| d |g\rangle = 0$. $\gamma_{A,1} |g\rangle$ and $\gamma_{B,N} |g\rangle$ have a relation

$$\langle g| \gamma_{A,1} \gamma_{B,N} |g\rangle = \left\langle g \left| \frac{1 - 2d^\dagger d}{2i} \right| g \right\rangle = \begin{cases} +\frac{i}{2} \text{ for } d^\dagger d = 1 \\ -\frac{i}{2} \text{ for } d^\dagger d = 0 \end{cases}. \tag{10.24}$$

It is determined by the number parity of $|g\rangle$. Thus, these two states are not independent. Therefore, the ground state of the Kitaev model are doubly degenerate, i.e., $|g\rangle$ and $d |g\rangle$, which have different parities: one is even and the other is odd. The double degeneracy reveals that the Kitaev model is topologically nontrivial. The solution is illustrated as in Fig. 10.2, which looks like the Affleck-Kohmoto-Lieb-Tasaki state for a spin-one system.

As an example, we solve a two-site problem. The Hamiltonian is

$$H = -t(c_x^\dagger c_{x+1} + c_{x+1}^\dagger c_x + c_x c_{x+1} + c_{x+1}^\dagger c_x^\dagger). \tag{10.25}$$

We have two eigenstates with even parity

$$\Psi_{e,\pm} = \frac{1}{\sqrt{2}} (1 \pm c_{x+1}^\dagger c_x^\dagger) |0\rangle \tag{10.26}$$

with the eigenvalues $\epsilon_{e,\pm} = \mp t$ and two eigenstates with odd parity

$$\Psi_{o,\pm} = \frac{1}{\sqrt{2}} (c_x^\dagger \pm c_{x+1}^\dagger) |0\rangle \tag{10.27}$$

with the eigenvalues $\epsilon_{e,\pm} = \mp t$.

In the language of the Majorana fermion operators,

$$H = -2it\gamma_{B,x}\gamma_{A,x+1} = +2td_x^\dagger d_x - t. \tag{10.28}$$

This Hamiltonian commutes with $\gamma_{A,x}$ and $\gamma_{B,x+1}$.

$$\gamma_{A,x}\Psi_{e,+} = \frac{i}{2}(c_x^\dagger + c_{x+1}^\dagger)\,|0\rangle\,, \tag{10.29}$$

$$\gamma_{B,x+1}\Psi_{e,+} = \frac{1}{2}(c_{x+1}^\dagger + c_x^\dagger)\,|0\rangle = -i\gamma_{A,x}\Psi_{e,+}. \tag{10.30}$$

Thus, these two states are identical up to a trivial phase factor.

10.2.3 Quasi-One-Dimensional Superconductor

The Kitaev model can be realized in a quasi-one-dimensional system. Recently, Potter and Lee [6] generalized the results to a quasi-one-dimensional system. They found that for a strip of two-dimensional p-wave superconductor when the width of strip is narrow enough such that the edge states at the two sides overlap in space, and open a finite energy gap as a result of finite size effect, the zero energy modes may exist at the two ends of the strip. The Majorana fermions of zero modes are quite robust against the disorder.

We consider a two-dimensional Kitaev model of spinless p-wave superconductors on a square lattice [7]:

$$H = \sum_{j=1}^{L}\sum_{\alpha=1}^{n}\Big[-\mu c_{j,\alpha}^\dagger c_{j,\alpha} - \Big(t c_{j,\alpha}^\dagger c_{j,\alpha+1} + \Delta c_{j,\alpha}c_{j,\alpha+1}$$
$$+t c_{j,\alpha}^\dagger c_{j+1,\alpha} + i\Delta c_{j,\alpha}c_{j+1,\alpha} + h.c.\Big)\Big], \tag{10.31}$$

where $c_{j,\alpha}^\dagger$ creates an electron on site (j,α), $t\ (> 0)$ is the hopping amplitude, μ is the chemical potential, Δ (for simplicity we take $\Delta > 0$) is the p-wave pairing amplitude. Here we consider a strip geometry in which the number of lattice sites is L along the x-axis direction and n along the y-axis direction (the sample width direction). Thus, the total number of lattice sites is $N = nL$. First, one introduces a periodic boundary condition along the x-axis direction, i.e., $c_{L+1,\alpha}^\dagger = c_{1,\alpha}^\dagger$, and uses the Fourier transform of the operator $c_{j,\alpha}^\dagger$:

$$c_{j,\alpha}^\dagger = \frac{1}{\sqrt{L}}\sum_q c_\alpha^\dagger(q)\,e^{-iqj}, \tag{10.32}$$

where $q = 2\pi l/L (l = 0, 1, \ldots, L - 1)$ is the wave vector along the x-axis, and $0 \leq q \leq 2\pi$. In terms of the new creation and annihilation operators $c_\alpha^\dagger (q)$ and $c_\alpha (q)$, the Hamiltonian (10.31) can be rewritten as

$$H = \sum_q \sum_{\alpha=1}^{n} \{ - (\mu + 2t \cos q) \, c_\alpha^\dagger (q) \, c_\alpha (q)$$

$$- \left[t c_\alpha^\dagger (q) \, c_{\alpha+1} (q) + |\Delta| \, c_\alpha (q) \, c_{\alpha+1} (-q) \right.$$

$$\left. + i \, |\Delta| \, e^{-iq} c_\alpha (q) \, c_\alpha (-q) + h.c. \right] \} . \tag{10.33}$$

Then, we define a set of the operators $\gamma_{2\alpha-1} (q)$ and $\gamma_{2\alpha} (q)$ as

$$\gamma_{2\alpha-1} (q) = i \left[c_\alpha^\dagger (-q) - c_\alpha (q) \right], \tag{10.34a}$$

$$\gamma_{2\alpha} (q) = c_\alpha^\dagger (-q) + c_\alpha (q), \tag{10.34b}$$

which satisfy the anticommutation relation $\{ \gamma_m^\dagger (q), \gamma_n (q') \} = 2\delta_{mn} \delta_{qq'}$ and $\gamma_m^\dagger (q) = \gamma_m (-q)$. In fact, $\gamma_m (0)$ is just a Majorana fermion operator due to $\gamma_m^\dagger (0) = \gamma_m (0)$. In the basis of the news operators $\gamma_{2\alpha-1} (q)$ and $\gamma_{2\alpha} (q)$, the Hamiltonian (10.33) has the following form:

$$H = i \frac{1}{4} \sum_q \sum_{\eta, \kappa} \gamma_\eta (-q) \, B_{\eta, \kappa} (q) \, \gamma_\kappa (q), \tag{10.35}$$

where the elements of the $2n \times 2n$ matrix $B (q)$ are given as

$$B_{2\alpha, 2\alpha} = -B_{2\alpha-1, 2\alpha-1} = -2i \, |\Delta| \sin q, \tag{10.36a}$$

$$B_{2\alpha, 2\alpha-1} = -B_{2\alpha-1, 2\alpha} = -\mu - 2t \cos q, \tag{10.36b}$$

$$B_{2\alpha, 2\alpha+1} = -B_{2\alpha+1, 2\alpha} = -t - |\Delta|, \tag{10.36c}$$

$$B_{2\alpha-1, 2\alpha+2} = -B_{2\alpha+2, 2\alpha-1} = t - |\Delta|, \tag{10.36d}$$

and all other elements are zero.

Here we will give the phase diagrams of the presence of Majorana end modes in quasi-one-dimensional p-wave superconductors by using topological arguments by Kitaev [5]. To this aim, we consider the $2n \times 2n$ matrix $B (q)$ in the Hamiltonian in Eq. (10.35). The matrix B is an antisymmetric matrix when q is equal to zero or π, such that we can calculate the Pfaffians $\text{Pf} B (0)$ and $\text{Pf} B (\pi)$. The topological property of the system described by the Hamiltonian in Eq. (10.35) is characterized by a Z_2 topological index (Majorana number) $\mathcal{M}$:

$$\mathcal{M} = \text{sgn} \left[\text{Pf} B (0) \right] \text{sgn} \left[\text{Pf} B (\pi) \right] = \pm 1, \tag{10.37}$$

where $+1$ corresponds to topologically trivial phases and -1 to topologically nontrivial states (i.e., the existence of zero-mode Majorana end states).

For the simplest case, there is only one lattice site along the y-axis direction (i.e., $n = 1$). This case is just the one-dimensional Kitaev model. Two 2×2 antisymmetric matrices are

$$B_{n=1} (0/\pi) = \begin{bmatrix} 0 & \mu \pm 2t \\ -(\mu \pm 2t) & 0 \end{bmatrix},$$
(10.38)

and $\mathrm{Pf} B_{n=1} (0/\pi) = \mu \pm 2t$, where "+" and "−" correspond to the cases of $q = 0$ and π, respectively. The Majorana number for the case of the strict one-dimensional limit is given:

$$\mathcal{M}_{n=1} = \mathrm{sgn}\,(\mu + 2t)\,\mathrm{sgn}\,(\mu - 2t) ;$$
(10.39)

thus, we have the topologically nontrivial condition

$$2\,|t| > |\mu|$$
(10.40)

with ($\Delta \neq 0$). The above Eq. (10.40) is just the result given by Kitaev [5], who demonstrated for a long open chain (in the limit of $L \to \infty$) there are zero-energy Majorana end states localized near per boundary point under the condition (10.40).

For the case of $n = 2$, the lattice site numbers along the y-axis direction are two. Two 4×4 antisymmetric matrices are

$$B_{n=2} (0/\pi) = \begin{bmatrix} 0 & \mu \pm 2t & 0 & t - |\Delta| \\ -(\mu \pm 2t) & 0 & -(t + |\Delta|) & 0 \\ 0 & t + |\Delta| & 0 & \mu \pm 2t \\ -(t - |\Delta|) & 0 & -(\mu \pm 2t) & 0 \end{bmatrix}.$$
(10.41)

The direct calculation yields the Pfaffians $\mathrm{Pf} B_{n=2} (0/\pi)$:

$$\mathrm{Pf} B_{n=2} (0/\pi) = (\mu \pm 2t)^2 + \Delta^2 - t^2.$$
(10.42)

For the larger lattice site numbers n (≥ 3), $\mathrm{Pf} B_n (0/\pi)$ can be also calculated analytically, and we obtain a recursion relation:

$$\mathrm{Pf} B_n (0/\pi) = a_{\pm} \mathrm{Pf} B_{n-1} (0/\pi) + b \mathrm{Pf} B_{n-2} (0/\pi),$$
(10.43)

where $a_{\pm} = \mu \pm 2t$ and $b = |\Delta|^2 - t^2$. We further solve Eq. (10.43) and give an analytic formula for $\mathrm{Pf} B_n (0/\pi)$:

$$\mathrm{Pf} B_n (0/\pi) = \frac{\left(r_1^{n+1} - r_2^{n+1}\right)}{\sqrt{a_{\pm}^2 + 4b}},$$
(10.44)

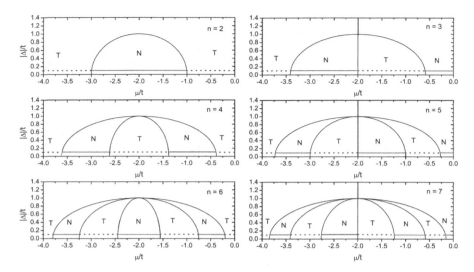

Fig. 10.3 Phase diagram for the quasi-1D p-wave superconductor model as a function of the p-wave pairing amplitude and chemical potential for lattice site numbers n along the y-axis direction. "N" denotes the topologically nontrivial region in the presence of zero-mode Majorana end states, and "T" denotes the topologically trivial region without zero-mode states. When $|\Delta|/t = 0.1$, the *solid (red) lines* and *dotted (blue) lines* guide the values of μ/t, corresponding to the topologically nontrivial and trivial phases, respectively (Adapted from [7])

where

$$r_1 = \frac{a_\pm + \sqrt{a_\pm^2 + 4b}}{2}, \quad r_2 = \frac{a_\pm - \sqrt{a_\pm^2 + 4b}}{2}. \tag{10.45}$$

According to the Pfaffians $\mathrm{Pf}B_n\,(0/\pi)$, one can compute $\mathcal{M}$ as a function of the physical parameters and then plot the phase diagram showing a sequence of topological phase transition for different lattice site numbers n. Figure 10.3 plots the phase diagram for the lattice site numbers n along the y-axis direction, respectively. The phase diagrams of this tight binding model have the symmetry on positive and negative μ values; thus, here we only plot on negative μ values because the other part on positive μ values is a mirror image. However, this $\mu \to -\mu$ symmetry is not generic to models with say, next-nearest-neighbor hopping or next-nearest-neighbor pairing.

10.3   Majorana Fermions in Topological Insulators

Fu and Kane proposed that as a superconducting proximity effect, the interface of the surface state of three-dimensional topological insulator and an s-wave superconductor resembles a spinless $p_x + ip_y$ superconductor, but does not break time reversal symmetry [8]. The system supports Majorana bound states at vortices. Suppose that an s-wave superconductor is deposited on the surface of topological insulator. Because of the proximity effect, Cooper pairs can tunnel into the surface states, which is described by the pairing potential $V = \Delta c^\dagger_{\mathbf{k},\uparrow} c^\dagger_{-\mathbf{k},\downarrow} + h.c.$ where $\Delta = \Delta_0 e^{i\phi}$. In the Nambu notation, $C^\dagger_{\mathbf{k}} = \{(c^\dagger_{\mathbf{k},\uparrow}, c^\dagger_{-\mathbf{k},\downarrow}), (c_{-\mathbf{k},\downarrow}, -c_{\mathbf{k},\uparrow})\}$, the surface states can then be described by

$$H = \frac{1}{2} \sum_{\mathbf{k}} C^\dagger_{\mathbf{k}} H_{\text{eff}}(\mathbf{k}) C_{\mathbf{k}}, \tag{10.46}$$

where

$$H_{\text{eff}} = -iv\tau_z\sigma \cdot \nabla - \mu\tau^z + \Delta_0(\tau_x \cos\phi - \tau_y \sin\phi) \tag{10.47}$$

where τ are Pauli matrices that mix the c and $c^\dagger$ blocks of C. The Hamiltonian has time reversal symmetry, $\Theta = i\sigma_y K$ (K is the complex conjugate operator), and the particle-hole symmetry, $\Xi = \sigma_y \tau_y K$. The energy spectrum is

$$E_{\mathbf{k}} = \pm\sqrt{(\pm vk - \mu)^2 + \Delta_0^2}. \tag{10.48}$$

For $\mu \gg \Delta_0$, the low-energy spectrum resembles that of a spinless $p_x + ip_y$ superconductor. Define $d_{\mathbf{k}} = (c_{\mathbf{k}\uparrow} + e^{i\theta_k} c_{\mathbf{k}\downarrow})^2$ for $\mathbf{k} = k_0(\cos\theta_k, \sin\theta_k)$ and $vk_0 \sim \mu$. The projected Hamiltonian is then

$$H_{\text{eff}} = \sum_{\mathbf{k}} (vk - \mu) d^\dagger_{\mathbf{k}} d_{\mathbf{k}} + \frac{\Delta_0}{2}(e^{i\theta_k} d^\dagger_{\mathbf{k}} d^\dagger_{-\mathbf{k}} + h.c). \tag{10.49}$$

This is identical to the one for p-wave pairing superconductor. Following the approach in p-wave superconductor, a half-quantum vortex in this system leads to a Majorana bound state.

10.4   Detection of Majorana Fermions

Consider two one-dimensional superconducting wires with Majorana end fermions connected at $x = 0$ to form a Josephson's junction. The effective Hamiltonian of the junction can be written in terms of two Majorana fermions at the two ends:

$$H_{\text{junction}} = 2i\,\Gamma(\phi)\gamma_{B,L}\gamma_{A,R} = \Gamma(\phi)\left(1 - 2d^\dagger_0 d_0\right) \tag{10.50}$$

where $\Gamma(\phi)$ is the coupling strength and is a function of the phase difference between the two superconductors, $\phi = \phi_R - \phi_L$. Suppose a gauge is chosen such that $\phi_L \to \phi_L + 2\pi$ and $\phi_R \to \phi_R$. In this gauge transformation, we have

Fig. 10.4 (*Top*) Schematic of
Josephson's junction of
Majorana fermions at two
superconductors with the
phase 0 and ϕ, respectively.
(*Bottom*) The energies of two
states with different parity as
a function of the phase ϕ

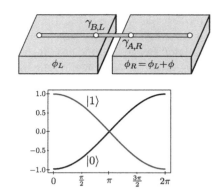

$$\gamma_{B,L} \;\rightarrow\; -\gamma_{B,L}, \tag{10.51a}$$

$$\gamma_{A,R} \;\rightarrow\; \gamma_{A,R}. \tag{10.51b}$$

As the Hamiltonian for superconductor is invariant under the gauge transformation,
we have a relation that $\Gamma(\phi) = -\Gamma(\phi + 2\pi)$. This condition shows that $\Gamma(\phi)$ is of
period 4π and crosses zero at $\phi = \pm\pi$.

Equation (10.50) shows that $\Gamma(\phi)$ and $-\Gamma(\phi)$ are the eigenvalues of H_{junction} and
$|0\rangle$ and $d_0^+ |0\rangle$ are the corresponding eigenstates. If the fermion parity is conserved
at the junction, this energy crossing is protected because $|0\rangle$ and $d_0^+ |0\rangle$ have
different fermion parities, and there is no transition from one state to the other when
ϕ equals π. Since the energy eigenvalues $\pm\Gamma(\phi)$ are periodic with 4π and there are
no transitions among the states with different fermion parity, the Josephson current,
which is given by

$$I_\pm = \pm\frac{2e}{h}\partial_\phi\Gamma(\phi), \tag{10.52}$$

is also 4π periodic. The observation of this 4π periodic Josephson current will be
an explicit evidence of Majorana fermions, and the sign of the current reveals the
fermion parity of the junction (Fig. 10.4).

There are a lot of other proposals to detect Majorana fermions, among which
tunneling spectroscopy is a direct method [11, 12]. Resonant tunneling into mid-gap
state produces a conductance of $2e^2/h$, while the conductance vanishes without
this state. Very recently, it was reported that the bound, mid-gap states at zero bias
voltage were observed in electric measurements on indium antimonide nanowires
contacted with one normal (gold) and one superconducting electrode [13]. Several
other groups also reported their experimental data to support existence of the
Majorana fermions [14, 15].

10.5   Sau-Lutchyn-Tewari-Das Sarma Model for Topological Superconductor

Sau et al. [9] proposed an idea to create Majorana fermions in a ferromagnetic insulator/semiconductor/s-wave superconductor hybrid system. They originally proved the existence of Majorana fermions in the setup by solving the vortex problem in the Bogoliubov-de Gennes equation. Alicea [10] found that the model is connected to a spinless $p + ip$ superconductor. Here, we prove that the system is actually equivalent to two spinless $p \pm ip$ superconductors, among which one is always topologically trivial and the other is possibly topologically nontrivial.

Consider first an isolated zinc-blende semiconductor quantum well grown along the (100) direction in the presence of a perpendicular Zeeman field. Assume the structural inversion asymmetry in the system, which generates Rashba spin-orbit coupling. The system can be modeled as a two-dimensional electron gas with Rashba spin-orbit coupling plus a perpendicular Zeeman field. The effective Hamiltonian reads

$$H_0 = \sum_{\mathbf{k},\sigma,\sigma'} c_{\mathbf{k},\sigma}^{\dagger} \left(\epsilon(k)\sigma_0 + \alpha(k_x\sigma_y - k_y\sigma_x) + V_Z\sigma_z \right)_{\sigma\sigma'} c_{\mathbf{k},\sigma'}, \tag{10.53}$$

where $\epsilon(k) = \frac{k^2}{2m} - \mu$. μ is the chemical potential and $\alpha (>0)$ is the Rashba spin-orbit coupling strength. Furthermore, consider the two-dimensional electron gas contacting an s-wave superconductor. Due to the proximity effect of superconductivity, an additional pairing potential is generated

$$V = \sum_{\mathbf{k}} \left(\Delta c_{\mathbf{k},\uparrow}^{\dagger} c_{-\mathbf{k},\downarrow}^{\dagger} + h.c. \right). \tag{10.54}$$

Thus, the total Hamiltonian for the electrons in quantum well becomes

$$H = H_0 + V. \tag{10.55}$$

To illustrate its connection to spinless p-wave superconductors, we first introduce a unitary transformation to diagonalize H_0,

$$\begin{pmatrix} c_{\mathbf{k},\uparrow} \\ c_{\mathbf{k},\downarrow} \end{pmatrix} = \begin{pmatrix} \cos\frac{\theta_{\mathbf{k}}}{2} & -e^{-i\varphi_{\mathbf{k}}}\sin\frac{\theta_{\mathbf{k}}}{2} \\ e^{i\varphi_{\mathbf{k}}}\sin\frac{\theta_{\mathbf{k}}}{2} & \cos\frac{\theta_{\mathbf{k}}}{2} \end{pmatrix} \begin{pmatrix} a_{\mathbf{k},+} \\ a_{\mathbf{k},-} \end{pmatrix}. \tag{10.56}$$

Consequently, H_0 is transformed to

$$H_0 = \sum_{\mathbf{k},\nu=\pm} (\epsilon(k) + \nu\Lambda_{\mathbf{k}}) \, a_{\mathbf{k},\nu}^{\dagger} a_{\mathbf{k},\nu}, \tag{10.57}$$

where $\Lambda_{\mathbf{k}} = \sqrt{V_Z^2 + \alpha^2 k^2}$. The parameters are determined by $\cos \theta_{\mathbf{k}} = V_Z / \Lambda_{\mathbf{k}}$, $\sin \theta_{\mathbf{k}} = \alpha k / \Lambda_{\mathbf{k}}$, $\cos \varphi_{\mathbf{k}} = -k_y / k$, and $\sin \varphi_{\mathbf{k}} = k_x / k$. After the transformation, the pairing potential V can be divided into two parts:

$$V = V_1 + V_2, \tag{10.58}$$

where

$$V_1 = \sum_{\mathbf{k}} \left(\Delta_{\mathbf{k},c} a_{\mathbf{k},+}^{\dagger} a_{-\mathbf{k},-}^{\dagger} + h.c. \right), \tag{10.59}$$

$$V_2 = -\frac{1}{2} \sum_{\mathbf{k},\nu} \left(\Delta_{\mathbf{k},\nu} a_{\mathbf{k},\nu}^{\dagger} a_{-\mathbf{k},\nu}^{\dagger} + h.c. \right), \tag{10.60}$$

$\Delta_{\mathbf{k},c} = \Delta \cos \theta_{\mathbf{k}}$, and $\Delta_{\mathbf{k},\pm} = \Delta e^{\mp i \varphi_{\mathbf{k}}} \sin \theta_{\mathbf{k}}$. Thus, $H_c = H_0 + V_1$ is equivalent to a s-wave superconductor with a "Zeeman" splitting, $\Lambda_{\mathbf{k}}$. We introduce the Bogoliubov transformation to diagonalize H_c:

$$\begin{pmatrix} a_{\mathbf{k},+} \\ a_{-\mathbf{k},-}^{\dagger} \end{pmatrix} = \begin{pmatrix} \cos \frac{\gamma_{\mathbf{k}}}{2} & -\sin \frac{\gamma_{\mathbf{k}}}{2} \\ \sin \frac{\gamma_{\mathbf{k}}}{2} & \cos \frac{\gamma_{\mathbf{k}}}{2} \end{pmatrix} \begin{pmatrix} b_{\mathbf{k},+} \\ b_{-\mathbf{k},-}^{\dagger} \end{pmatrix}, \tag{10.61}$$

where $\cos \gamma_{\mathbf{k}} = \epsilon(k) / \sqrt{\epsilon(k)^2 + \Delta_{\mathbf{k},c}^2}$ and $\sin \gamma_{\mathbf{k}} = \Delta_{\mathbf{k},c} / \sqrt{\epsilon(k)^2 + \Delta_{\mathbf{k},c}^2}$. As a result,

$$H_c = \sum_{\mathbf{k},\nu=\pm} \left(\sqrt{\epsilon(k)^2 + \Delta_{\mathbf{k},c}^2} + \nu \Lambda_{\mathbf{k}} \right) b_{\mathbf{k},\nu}^{\dagger} b_{\mathbf{k},\nu} \tag{10.62}$$

up to a constant. Meanwhile, the pairing potential V_2 has the form

$$V_2 = -\frac{1}{2} \sum_{\mathbf{k},\nu} \left(\Delta_{\mathbf{k},\nu} b_{\mathbf{k},\nu}^{\dagger} b_{-\mathbf{k},\nu}^{\dagger} + h.c. \right). \tag{10.63}$$

In the language of $b_{\mathbf{k},\pm}^{\dagger}$ and $b_{\mathbf{k},\pm}$, the pairing potential V_2 consists of two pairing potentials between the same types of the particles. The particles with $\nu = +$ and $\nu = -$ are decoupled completely. Therefore, the total Hamiltonian is reduced to

$$H = \sum_{\mathbf{k},\nu=\pm} \left[\left(\sqrt{\epsilon(k)^2 + \Delta_{\mathbf{k},c}^2} + \nu \Lambda_{\mathbf{k}} \right) b_{\mathbf{k},\nu}^{\dagger} b_{\mathbf{k},\nu} - \frac{1}{2} \sum_{\mathbf{k},\nu} \left(\Delta_{\mathbf{k},\nu} b_{\mathbf{k},\nu}^{\dagger} b_{-\mathbf{k},\nu}^{\dagger} + h.c. \right) \right]. \tag{10.64}$$

The order parameters are

$$\Delta_{\mathbf{k},\nu} = -\frac{\alpha \Delta}{\sqrt{V_Z^2 + \alpha^2 k^2}} \left(k_y + i \nu k_x \right), \tag{10.65}$$

which obey $p \pm ip$ symmetry. Thus, the effective model consists of two different types of spinless $(p_x \pm ip_y)$-wave pairing superconductors. By introducing a Nambu spinor, $\psi_{\mathbf{k},\nu}^{\dagger} = \left(b_{\mathbf{k},\nu}^{\dagger}, b_{-\mathbf{k},\nu} \right)$, the total Hamiltonian has the form

$$
H = \frac{1}{2} \sum_{\mathbf{k},\nu=\pm} \psi_{\mathbf{k},\nu}^{\dagger} \left[\frac{\alpha \Delta}{\Lambda_{\mathbf{k}}} \left(k_y \sigma_x - \nu k_x \sigma_y \right) + \left(\sqrt{\epsilon(k)^2 + \Delta_{\mathbf{k},c}^2} + \nu \Lambda_{\mathbf{k}} \right) \sigma_z \right] \psi_{\mathbf{k},\nu}
$$

(10.66)

by ignoring a constant.

The equation can be reduced to two modified Dirac equations near $k = 0$. In each type of the superconductor, the factor $\alpha \Delta / V_Z$, which is equivalent to the velocity in the modified Dirac equation, plays a role in coupling the two bands. This fact illustrates that Rashba spin-orbit coupling, the pairing potential, and the Zeeman field combine together to form three indispensable ingredients to realize a topological superconductor. For the particles of $\nu = +$, the spectrum is

$$
\epsilon_+ = \sqrt{ \frac{\alpha^2 \Delta^2}{V_Z^2 + \alpha^2 k^2} k^2 + \left(\sqrt{V_Z^2 + \alpha^2 k^2} + \sqrt{\epsilon(k)^2 + \frac{V_Z^2 \Delta^2}{V_Z^2 + \alpha^2 k^2}} \right)^2 }.
$$

(10.67)

The gap between the particle and hole bands is always positive and never close if $\Delta \neq 0$. The Chern number (see Sect. A.2) is equal to zero. Thus, it is always topologically trivial. For the particles of $\nu = -$, the spectrum is

$$
\epsilon_- = \sqrt{ \frac{\alpha^2 \Delta^2 k^2}{V_Z^2 + \alpha^2 k^2} + \left(\sqrt{V_Z^2 + \alpha^2 k^2} - \sqrt{\epsilon(k)^2 + \frac{V_Z^2 \Delta^2}{V_Z^2 + \alpha^2 k^2}} \right)^2 }.
$$

(10.68)

The gap can be closed only at $k = 0$ if $\Delta \neq 0$. Near the point, it follows from (10.66) that the gap can be either positive or negative,

$$
\Delta_{GAP}^+ = \left(\Lambda - \sqrt{\epsilon(k)^2 + \Delta_{\mathbf{k},c}^2} \right)_{k=0} = |V_Z| - \sqrt{\mu^2 + \Delta^2}.
$$

(10.69)

The sign of the gap itself does not determine the topology of the band structure. However, the sign change demonstrates that a topological quantum phase transition can occur near $|V_Z| = \sqrt{\mu^2 + \Delta^2}$. The Chern number for the hole band of $b_{\mathbf{k},-}$ can be calculated explicitly,

$$
n_c = \frac{1}{2} \left[\mathrm{sgn} \left(\sqrt{\mu^2 + \Delta^2} - |V_Z| \right) - 1 \right].
$$

(10.70)

Therefore, it is topologically nontrivial if $\sqrt{\mu^2 + \Delta^2} < |V_Z|$, while it is topologically trivial if $\sqrt{\mu^2 + \Delta^2} > |V_Z|$.

10.6   Non-Abelian Statistics and Topological Quantum Computing

If the overall phase of the superconducting gap shifts by ϕ, it is equivalent to rotating the creation and annihilation operators of electron by $\phi/2$: $c \rightarrow e^{i\phi/2}c$ and $c^\dagger \rightarrow e^{-i\phi/2}c^\dagger$. The solution for the Majorana fermion $\gamma = uc^\dagger + u^*c \rightarrow ue^{-i\phi/2}c^\dagger + u^*e^{i\phi/2}c$. If the phase of the order parameter is changed by 2π, the Majorana fermion in the vortex changes its sign: $\gamma \rightarrow -\gamma$. Let us fix the initial positions of vortices. Permutations of the vortices may form a braid group B_{2n}, which is generated by the elementary interchange T_i of neighboring vortices [4].

Under the action T_i:

$$\gamma_i \rightarrow \gamma_{i+1}, \tag{10.71a}$$

$$\gamma_{i+1} \rightarrow -\gamma_i, \tag{10.71b}$$

$$\gamma_j \rightarrow \gamma_j \tag{10.71c}$$

for $j \neq i$ and $j \neq i + 1$. This action obeys the commutation relations:

$$T_i T_j = T_j T_i, \text{ for } |i - j| > 1, \tag{10.72a}$$

$$T_i T_j T_i = T_j T_i T_j, \text{ for } |i - j| = 1, \tag{10.72b}$$

which is for the braid group. The expression for this action is

$$\tau(T_i) = \exp\left(\frac{\pi}{2}\gamma_{i+1}\gamma_i\right) = \exp\left(-i\frac{\pi}{4}P\right)$$

$$= \cos\frac{\pi}{4} - iP\sin\frac{\pi}{4} = \frac{1}{\sqrt{2}}(1 + 2\gamma_{i+1}\gamma_i) \tag{10.73}$$

where P is the parity operator and $P^2 = P$. Thus, the Majorana fermions associated with a quantum vortex obey non-Abelian statistics.

In the case of four vortices, the four Majorana fermions combine into two complex fermions $c_1 = \frac{1}{\sqrt{2}}(\gamma_1 + i\gamma_2)$ and $c_2 = \frac{1}{\sqrt{2}}(\gamma_3 + i\gamma_4)$. The ground state is fourfold degenerated, and the three generators T_1, T_2, and T_3 of the braid group are represented by

$$\tau(T_1) = \exp\left(\frac{\pi}{2}\gamma_2\gamma_1\right), \tag{10.74a}$$

$$\tau(T_2) = \exp\left(\frac{\pi}{2}\gamma_3\gamma_2\right), \tag{10.74b}$$

$$\tau(T_3) = \exp\left(\frac{\pi}{2}\gamma_4\gamma_3\right). \tag{10.74c}$$

One may write the operators in a matrix form in the basis $\{|0\rangle, c_1^\dagger |0\rangle, c_2^\dagger |0\rangle, c_1^\dagger c_2^\dagger |0\rangle\}$.

$$\tau(T_1) = \exp\left(\frac{\pi}{2}\gamma_2\gamma_1\right) = \begin{pmatrix} e^{-i\frac{\pi}{4}} & 0 & 0 & 0 \\ 0 & e^{+i\frac{\pi}{4}} & 0 & 0 \\ 0 & 0 & e^{-i\frac{\pi}{4}} & 0 \\ 0 & 0 & 0 & e^{+i\frac{\pi}{4}} \end{pmatrix}, \tag{10.75}$$

$$\tau(T_2) = \exp\left(\frac{\pi}{2}\gamma_3\gamma_2\right) = \frac{1}{\sqrt{2}}\begin{pmatrix} 1 & 0 & 0 & -i \\ 0 & 1 & -i & 0 \\ 0 & -i & 1 & 0 \\ -i & 0 & 0 & 1 \end{pmatrix}, \tag{10.76}$$

$$\tau(T_3) = \exp\left(\frac{\pi}{2}\gamma_4\gamma_3\right) = \begin{pmatrix} e^{-i\frac{\pi}{4}} & 0 & 0 & 0 \\ 0 & e^{-i\frac{\pi}{4}} & 0 & 0 \\ 0 & 0 & e^{+i\frac{\pi}{4}} & 0 \\ 0 & 0 & 0 & e^{+i\frac{\pi}{4}} \end{pmatrix}. \tag{10.77}$$

A quantum computation consists of three steps:

1. Create: if a pair of i, j of vortices is created, they will be in the ground state $|0_{ij}\rangle$ with no extra quasiparticle excitations. Creating N pairs initialize the system.
2. Braid: adiabatically rearranging the vortices modifies the state and performs a quantum computation.
3. Measure: bringing vortices i and j back together allows the quantum state associated with each pair to be measured. $|0_{ij}\rangle$ will be distinguished by the presence or absence of extra fermionic quasiparticle associated with the pair.

Majorana fermions might provide the basic elements for a quantum computer. This is the motivation behind the search of Majorana fermions in condensed matter systems.

10.7 Further Reading

- A.Yu. Kitaev, Unpaired Majorana fermions in quantum wires. Physics-Uspekhi **44**, 131 (2001)
- L. Fu, C.L. Kane, Superconducting proximity effect and Majorana fermions at the surface of a topological insulator. Phys. Rev. Lett. **100**, 096407 (2008)
- F. Wilczek, Majorana returns. Nat. Phys. **5**, 614 (2009)
- J. Alicea, New directions in the pursuit of Majorana fermions in solid state systems. arXiv: 1202.1293v1[cond-mat.supr-con]
- C.W.J. Beenakker, Search for Majorana fermions in superconductors. arXiv: 1112.1950v2 [cond-mat.mes-hall]

References

1. P.A.M. Dirac, *Principles of Quantum Mechanics*, 4th edn. (Clarendon, Oxford, 1982)
2. E. Majorana, Nuovo Cimento **5**, 171 (1937)
3. F. Wilczek, Nat. Phys. **5**, 614 (2009)
4. D.A. Ivanov, Phys. Rev. Lett. **86**, 268 (2001)
5. A.Yu Kitaev, Phys. Usp. **44**, 131 (2001)
6. A.C. Potter, P.A. Lee, Phys. Rev. Lett. **105**, 227003 (2010)
7. B. Zhou, S.Q. Shen, Phys. Rev. B **84**, 054532 (2011)
8. L. Fu, C.L. Kane, Phys. Rev. Lett. **100**, 096407 (2008)
9. J.D. Sau, R.M. Lutchyn, S. Tewari, S. Das Sarma, Phys. Rev. Lett. **104**, 040502 (2010)
10. J. Alicea, Phys. Rev. B **81**, 125318 (2010).
11. C.J. Bolech, E. Demler, Phys. Rev. Lett. **98**, 237002 (2007)
12. K.T. Law, P.A. Lee, T.K. Ng, Phys. Rev. Lett. **103**, 237001 (2009)
13. V. Mourik, K. Zuo, S.M. Frolov, S.R. Plissard, E.P.A.M. Bakkers, L.P. Kouwenhoven, Science **336**, 6084 (2012)
14. J.R. Williams, A.J. Bestwick, P. Gallagher, S.S. Hong, Y. Cui, A.S. Bleich, J.G. Analytis, I.R. Fisher, D. Goldhaber-Gordon. Phys. Rev. Lett. **109**, 056803 (2012)
15. M.T. Deng, C.L. Yu, G.Y. Huang, M. Larsson, P. Caroff, H.Q. Xu. arXiv: 1204.4130v1 (cond-mat.mes-hall)

Chapter 11
Topological Anderson Insulator

Abstract Topological Anderson insulator is a distinct type of topological insulator, which is induced by the disorders. Its key difference from the conventional topological insulators is that its Fermi energy lies within a so-called mobility gap instead of a "real" band gap. The robustness of the edge or surface states is protected by the mobility gap.

Keywords Topological Anderson insulator • Quantized anomalous Hall effect • Band gap • Mobility gap

11.1 Band Structure and Edge States

We start with a two-dimensional ferromagnetic metal with strong spin-orbit coupling:

$$h(\mathbf{k}) = \epsilon(k) + \mathbf{d}(\mathbf{k}) \cdot \sigma, \tag{11.1}$$

where $\mathbf{d}(\mathbf{k}) = (Ak_x, Ak_y, M - Bk^2)$ and $\epsilon(k) = C - Dk^2$ with A, B, C, and D being sample-specific parameters. This is a modified Dirac equation plus an additional term $\epsilon(k)$, which breaks the symmetry between the conduction and valence bands. In order to keep the band gap open, we require that $B^2 > D^2$. In this case the Chern number for this model is given by

$$n_c = -\frac{1}{2} \left[\text{sgn}(M) + \text{sgn}(B) \right]. \tag{11.2}$$

For positive B, it tells that the sign change of M signifies a topological quantum phase transition between a conventional insulating phase ($M < 0$ and $n_c = 0$) and a topological quantum phase ($M > 0$ and $n_c = 1$). Nonzero Chern number indicates that the Hall conductance is quantized, $\sigma_H = n_c e^2 / h$. Thus, the existence of the additional term $\epsilon(k)$ does not affect the Chern number once the band gap keeps open.

In an infinite-length strip with open lateral boundary conditions, the solution of the two-band model $\mathcal{H}\Psi = E\Psi$ is given by Zhou et al. [1]

$$\Psi(k_x, y) = (\mu_+ e^{\alpha y} + \mu_- e^{-\alpha y} + \nu_+ e^{\beta y} + \nu_- e^{-\beta y}), \qquad (11.3)$$

$\mu_\pm$ and $\nu_\pm$ are two-component k_x-dependent coefficients and α, β are determined self-consistently by the following set of equations:

$$\alpha^2 = k_x^2 + F - \sqrt{F^2 - \frac{M^2 - E^2}{B^2 - D^2}}, \qquad (11.4)$$

$$\beta^2 = k_x^2 + F + \sqrt{F^2 - \frac{M^2 - E^2}{B^2 - D^2}}, \qquad (11.5)$$

$$E_\alpha^2 \beta^2 + E_\beta^2 \alpha^2 - \gamma E_\alpha E_\beta \alpha \beta = k_x^2 (E_\alpha - E_\beta)^2. \qquad (11.6)$$

Here, we have

$$F = \frac{A^2 - 2(MB + ED)}{2(B^2 - D^2)}, \qquad (11.7)$$

$$E_\alpha = E - M + (B + D)(k_x^2 - \alpha^2), \qquad (11.8)$$

$$E_\beta = E - M + (B + D)(k_x^2 - \beta^2), \qquad (11.9)$$

$$\gamma = \frac{\tanh \frac{\alpha L_y}{2}}{\tanh \frac{\beta L_y}{2}} + \frac{\tanh \frac{\beta L_y}{2}}{\tanh \frac{\alpha L_y}{2}}, \qquad (11.10)$$

and L_y is the width of the strip. We take the Dirichlet boundary condition at $y = \pm L_y/2$:

$$\Psi\left(k_x, y = \pm \frac{L_y}{2}\right) = 0. \qquad (11.11)$$

The solutions of this set of equations naturally contain both helical edge states ($\alpha^2 < 0$) and bulk states ($\alpha^2 > 0$), which are shown in Fig. 11.1 for three cases $M < 0$, $M = 0$, and $M > 0$. The edge states (red lines in Fig. 11.1) are seen beyond the bulk gap for all cases, up to an M-dependent maximum energy. When $M < 0$, the edge states cross the bulk gap producing a quantum Hall effect. At $M = 0$, the edge states exist only in conjunction with the lower band, terminating at the Dirac point. For $M > 0$, there are no edge states in the gap, producing a conventional insulator, but the edge states may coexist with the valence band. Appearance of edge state is a key feature of this model even for a normal band structure although these states mix with the bulk states.

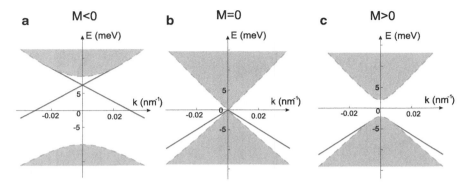

Fig. 11.1 Band structure of HgTe/CdTe quantum wells in a geometry of stripe with finite width. (**a**) The "inverted" band structure case with $M = -10$ meV. Edge states (*red solid line*) cross the bulk band gap and merge into bulk states (*gray area*) at a maximum energy in the *upper* band. The *dashed line* mark the boundary of bulk states. (**b**) The transition point between an inverted band structure and a "normal" band structure with $M = 0$ meV. (**c**) The normal band structure with $M = 2$ meV. In all figures, the strip width L_y is set to $100\,\mu m$. The sample-specific parameters are fixed to be $A = 364.5$ meV nm, $B = -686$ meV nm^2, $C = 0$, and $D = -512$ meV nm^2 (Adapted from [2])

11.2   Quantized Anomalous Hall Effect

For numerical simulation, we take the tight binding approximation on a square lattice, and the Hamiltonian has the form

$$\mathbf{d(k)} = \left(A \sin k_x,\ A \sin k_y,\ M - 4B \sin^2 \frac{k_x^2}{2} - 4B \sin^2 \frac{k_y^2}{2} \right) \qquad (11.12)$$

for the periodic boundary condition. In the lattice space, after performing the Fourier transformation, we have a lattice model as in Chap. 3.

The most surprising aspect revealed by numerical calculation is the appearance of quantized anomalous conductance at a large disorder for situation when the clean limit system is a metal without preexisting edge state. We study transport as a function of disorder, with the Fermi energy varying through all regions of the band structure. For this purpose, disorders are introduced through random on-site energy with a uniform distribution within $[-W/2, W/2]$. The conductance of disordered strips of width L_y and length L_x was calculated in a two-terminal setup using the Landauer-Büttiker formalism [3, 4]. The conductance G as a function of disorder strength W is plotted in Fig. 11.2. Furthermore, the conductance was scaled with the width of the strip. Figure 11.2 shows the calculated conductance of a strip as a function of its width L_y. In the region before the quantized anomalous conductance plateau is reached, the scaled conductance GL_x/L_y, or conductivity, is independent of width, as shown in the inset of Fig. 11.2, which implies bulk transport. Within

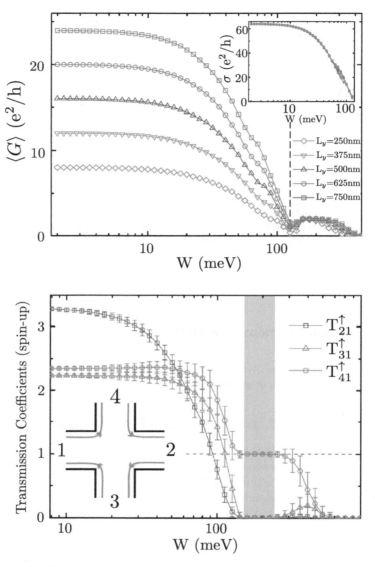

Fig. 11.2 (*Left*): Width-dependence of the conductance in disordered strips with several values of strip width L_y and a length $L_x = 2,000$ nm. In the inset, the conductance traces prior to the quantum anomalous Hall phase (left-handside of the *dashed line*) are scaled with the width of the strips as $\sigma = GL_x/L_y$. The formation of the edge states is indicated by the presence of conductance quantization e^2/h. In this figure, $M = 2$ meV and the Fermi energy $E_f = 20$ meV. (*Right*): Three independent spin-resolved transmission coefficients, $T_{21}^\uparrow$, $T_{31}^\uparrow$, and $T_{41}^\uparrow$, are plotted as functions of disorder strength W. Standard deviations of the transmission coefficients for 1,000 samples are shown as the error bars. In the shadowed range of disorder strength, all bulk states are localized and only chiral edge states exist, which is schematically shown in the inset (for spin-up component only). The width of leads is 500 nm and $M = 1$ meV and $E_f = 20$ meV (Adapted from [2])

the quantized plateau, absence of such scaling indicates a total suppression of the bulk conduction, thus confirming presence of conducting edge states in an otherwise localized system.

We further examine the picture of edge-state transport in a four-terminal cross-bar setup by calculating the spin-resolved transmission coefficients T_{pq} between each ordered pair of leads p and q ($= 1, 2, 3, 4$). Three independent coefficients, T_{21}, T_{31}, and T_{41}, are shown in Fig. 11.2 as functions of the disorder strength inside the cross region. The shadowed area marks the appearance of quantized plateau, where $\langle T_{41} \rangle = 1$, $\langle T_{21} \rangle = \langle T_{31} \rangle = 0$, and all transmission coefficients exhibit vanishingly small fluctuations. From symmetry, it follows that $\langle T_{41} \rangle = \langle T_{24} \rangle = \langle T_{32} \rangle = \langle T_{13} \rangle \to 1$, and all other coefficients are vanishing small. These facts are easily understood from the presence of a chiral edge state. Two consequences of this chiral edge state transport are a vanishing diagonal conductance $G_{xx} = (T_{21} - T_{12})e^2/h = 0$ and a quantized Hall conductance $G_{xy} = (T_{41} - T_{42})e^2/h = e^2/h$, analogous to Haldane's model for the integer quantum Hall effect with parity anomaly [5]. The quantized Hall conductance G_{xy} reveals that the topologically invariant Chern number of this state is equal to one. Thus, this is a disorder-induced quantum anomalous Hall effect.

A noncommutative Chern number can be defined in disordered system. Prodan [6] did a series of calculation for the disordered system and found that the Chern number takes a quantized value ± 1.

11.3 Topological Anderson Insulator

Now we are ready for topological Anderson insulator, which does not break the time reversal symmetry. The effective Hamiltonian for a clean bulk HgTe/CdTe quantum well is given by Bernevig et al. [7]

$$\mathcal{H}(k) = \begin{pmatrix} h(k) & 0 \\ 0 & h^*(-k) \end{pmatrix}, \tag{11.13}$$

where $h(k)$ has the identical form of the 2×2 Hamiltonian for two-dimensional ferromagnet with spin-orbit coupling. This 4×4 model is a combination of $h(k)$ and $h^*(-k)$ which is the time reversal counterpart of $h(k)$. The model is equivalent to the two-dimensional modified Dirac model in Eq. (2.32) with an additional kinetic energy term $\epsilon(k)$. When $h(k)$ contributes a Hall conductance e^2/h, its time reversal counterpart $h^*(-k)$ will also contribute a quantum Hall conductance, but with an opposite sign, $-e^2/h$. As a result, the total Hall conductance in this system is always equal to zero. Both $h(k)$ and $h^*(-k)$ produce a chiral edge state: electrons in one edge state of $h(k)$ are moving in one direction, and electrons in another edge state are moving in opposite direction. The electron spins in the two states are connected by the time reversal operation and must be antiparallel. Therefore, this is a quantum spin Hall effect in $\mathcal{H}(k)$.

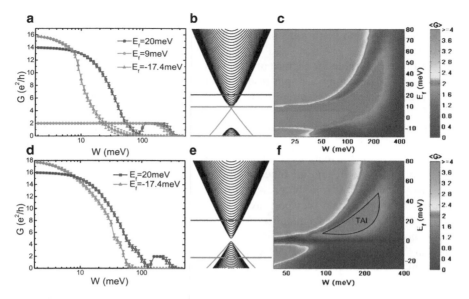

Fig. 11.3 Conductance of disordered strips of HgTe/CdTe quantum wells. The *upper* panels (**a**)–
(**c**) show results for an quantum well "inverted" with $M = -10$ meV, and the *lower* panels
(**d**)–(**f**) for a "normal" quantum well with $M = 1$ meV. (**a**) The conductance G as a function
of disorder strength W at three values of Fermi energy. The error bars show standard deviation
of the conductance for 1,000 samples. (**b**) Band structure calculated with the tight-binding model.
Its vertical scale (energy) is same as in (**c**) and the *horizontal lines* correspond to the values of
Fermi energy considered in (**a**). (**c**) Phase diagram showing the conductance G as a function of
both disorder strength W and Fermi energy E_f. The panels (**d**)–(**f**) are same as (**a**)–(**c**), but for
$M > 0$. The TAI phase regime is labeled. In all figures, the strip width L_y is set to 500 nm; the
length L_x is 5,000 nm in (**a**) and (**d**) and 2,000 nm in (**c**) and (**f**) (Adapted from [2])

The calculated behavior conforms to the qualitative expectation for certain
situations. For Fermi level in the lower band, for both $M < 0$ and $M > 0$, an
ordinary Anderson insulator results when the clean limit metal is disordered (green
lines in Fig. 11.3a, d). The conductance in this case decays to zero at disorder
strength around 100 meV, which is about five times of the conventional hopping
energy between nearest neighboring sites $t = -D/a^2 \approx 20.5$ meV, and much
larger than the clean-limit bulk band gap $E_g = 2|M| = 20$ meV. Here $a = 5$
nm is the lattice spacing of the tight binding model. The topological insulator (red
line in Fig. 11.3a) is robust and requires a strong disorder before it eventually yields
to a localized state. This is expected as a result of the absence of backscattering in a
topological insulator when time reversal symmetry is preserved [8].

The most surprising aspect revealed by our calculations is the appearance of
anomalous conductance plateaus at large disorder for situations when the clean limit
system is a metal without preexisting edge states. See, for example, the blue lines in
Fig. 11.3a ($M < 0$) and Fig. 11.3d ($M > 0$). The anomalous plateau is formed after

the usual metal-insulator transition in such a system. The conductance fluctuations (the error bar in Fig. 11.3a, d) are vanishingly small on the plateaus; at the same time the Fano factor drops to nearly zero indicating the onset of dissipationless transport in this system, even though the disorder strength in this scenario can be as large as several 100 meV. This state is termed topological Anderson insulator. The quantized conductance cannot be attributed to the relative robustness of edge states against disorder, because it occurs for cases in which no edge states exist in the clean limit. The irrelevance of the clean limit edge states to this physics is further evidenced from the fact that no anomalous disorder-induced plateaus are seen for the clean limit metal for which bulk and edge states coexist; those exhibit a direct transition into an ordinary Anderson insulator.

The nature of topological Anderson insulator is further clarified by the phase diagrams shown in Fig. 11.3c for $M < 0$ and in Fig. 11.3f for $M > 0$. For $M < 0$, the quantized conductance region (green area) of the topological Anderson insulator in the upper band is connected continuously with the quantized conductance area of the topological insulator phase of the clean-limit. One cannot distinguish between these two phases by the conductance value. When $M > 0$, however, the anomalous conductance plateau occurs in the highlighted green island labeled TAI (topological Anderson insulator), surrounded by an ordinary Anderson insulator. No plateau is seen for energies in the gap, where a trivial insulator is expected. The topology of the topological Anderson insulator and the absence of preexisting edge states in the clean limit demonstrate that the topological Anderson insulator owes its existence fundamentally to disorder.

The existence of topological Anderson insulator has been confirmed by several independent groups. As a new type of topological insulator, topological Anderson insulator exists even in three dimensions [9]. To confirm the genuine three-dimensional nature of the topological Anderson insulator, Guo et al. probed for the Witten effect in their three-dimensional model. According to Witten, a magnetic monopole in a media could bind electric charge $-e(n + \frac{1}{2})$ with an integer n. They found that a half charge is bound to a monopole in three-dimensional topological Anderson insulator by numerical calculation.

11.4 Effective Medium Theory for Topological Anderson Insulator

Groth et al. [10] proposed an effective medium theory to explain the disorder-induced transition from a conventional metal to a topological Anderson insulator. Consider a scalar short-ranged potential for the disorder: $V(\mathbf{r}) = V_0 \sum_i \delta(\mathbf{r} - \mathbf{R}_i)$, where V_0 is the strength of disorder. The retarded Green's function can be written as

$$G^R(k, E, \Sigma^R) = (E - h(k) - \Sigma^R)^{-1}. \tag{11.14}$$

Here the self-energy Σ^R is defined by

$$(E_F - h(k) - \Sigma^R)^{-1} = \left\langle \frac{1}{E_F - h(k) - V(r)} \right\rangle \qquad (11.15)$$

with $\langle \cdots \rangle$ the disorder average. The self-energy can be expanded in terms of the Pauli matrices: $\Sigma^R = \sum_{i=0,x,y,z} \Sigma_i \sigma_i$. Thus, in the effective Hamiltonian, $H_{\text{eff}} = h(k) + \Sigma^R$, the renormalized parameters are given by

$$\tilde{M} = M + \lim_{k \to 0} \text{Re} \Sigma_z, \qquad (11.16a)$$

$$\tilde{E}_F = E_F - \lim_{k \to 0} \text{Re} \Sigma_0. \qquad (11.16b)$$

The phase boundary of the topological Anderson insulator is at $\tilde{M} = 0$, while the Fermi level enters the negative band gap when $\tilde{E}_F = -\tilde{M}$. In the Born approximation, the self-energy is given by the integral equation

$$\Sigma^R = \frac{1}{12} \left(\frac{a}{2\pi} \right)^2 V_0^2 \int_{BZ} \frac{d\mathbf{k}}{(2\pi)^2} G^R(k, E_F + i0^+, \Sigma^R). \qquad (11.17)$$

where the integral runs over the first Brillouin zone (BZ).

An approximate solution can be derived in a closed form:

$$\tilde{M} = M + \frac{V_0^2 a^2}{48\pi\hbar^2} \frac{B}{B^2 - D^2} \ln \left| \frac{B^2 - D^2}{E_F^2 - M^2} \left(\frac{\pi\hbar}{a} \right)^2 \right|, \qquad (11.18a)$$

$$\tilde{E}_F = E_F + \frac{V_0^2 a^2}{48\pi\hbar^2} \frac{D}{B^2 - D^2} \ln \left| \frac{B^2 - D^2}{E_F^2 - M^2} \left(\frac{\pi\hbar}{a} \right)^2 \right|. \qquad (11.18b)$$

In the clean limit, if M and B have different signs, say $B > 0$ but $M < 0$, the system is a conventional metal. The modification of $\delta M = \tilde{M} - M$ is positive provided $B^2 > D^2$ which is the condition for the gap opening between the conduction and valence bands. This will change a negative M into a positive $\tilde{M}$, leading to a quantum phase transition.

This theory describes very well the transition from a metal into topological Anderson insulator in a weak disorder, but fails to predict the transition from topological Anderson insulator to Anderson insulator in an even stronger disorder.

11.5 Band Gap or Mobility Gap

The edge or surface states in topological Anderson insulator are expected to be protected by the mobility gap instead of the band gap as in topological (band) insulator. In this section, by doing statistics on the local density of states (DOS), a function of energy, it is possible to identify which states are localized and which states are extended. The kernel polynomial method is a powerful method for evaluating spectrum properties [11–13].

There are two distinct average DOS in a disordered calculation. The average DOS is defined as the algebraic average of the local DOS,

$$\rho_{av} = \langle \rho_i(E) \rangle ; \tag{11.19}$$

the typical DOS is defined as the geometric average of the local DOS,

$$\rho_{typ} = \exp[\langle \ln(\rho_i(E)) \rangle]. \tag{11.20}$$

When electron states are extended, the DOS distribution is almost uniform in the space, and thus, there should be not much difference between the two definitions. However, when electron states are localized, the DOS is high near some sites, but almost vanishes on the others. Thus, we expect significant ratio between the two types of DOS [14].

We can take a lattice sample of periodic boundary condition on both x- and y-direction (i.e., a torus) and do the statistics of Eqs. (11.19) and (11.20). The upper block of the Hamiltonian in Eq. (11.13) is used, and we take $A = B = 1, C = D = 0$ such that the electron-hole symmetry is recovered. $M = 0.2$ such that the system is initially a trivial band insulator. The result is plotted in Fig. 11.4 with increasing disorder strength from Fig. 11.4a–e.

The mass renormalization phenomenon proposed by Groth et al. [10] is confirmed in the weak disorder regime, where initially a band gap in Fig. 11.4a is clearly seen but is gradually closed in Fig. 11.4b as the disorder increases. As is seen from the ratio ρ_{typ}/ρ_{av}, at the strong disorder regime, we can observe two extended states at $E = \pm 1$ in Fig. 11.4c, which indicates that the system is topologically nontrivial. It is noted that no band gap opens again for a stronger disorder while the mobility gap opens to separate the two extended states. As the disorder further increases, these two extended states move toward each other and finally collide and disappear in Fig. 11.4d. Finally, all the states become localized. This phenomenon can be identified as the levitation and pair annihilation. Levitation and pair annihilation is the hallmark of any extended states carrying topological numbers. Such states are stable against disorder until those with opposite topological numbers collide with each other and become trivial when disorder strength is increased. In the disorder-induced nontrivial Hamiltonian $h(k)$, there exist the gapless edge states between the two extended states, which is the origin of topological Anderson insulator.

11.6 Summary

Topological Anderson insulator is distinct from the conventional topological insulator, or topological band insulator. We find that there exists a mobility gap instead of the band gap in the system. From the point of view of time reversal symmetry,

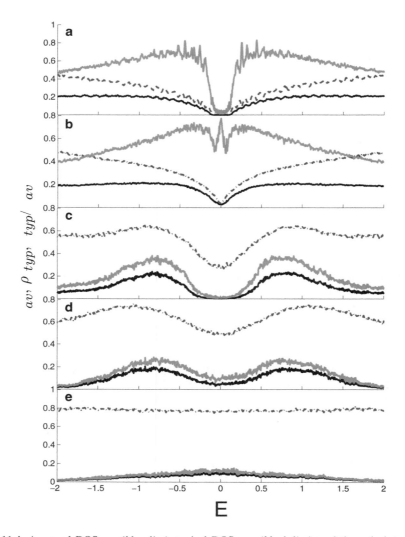

Fig. 11.4 Averaged DOS ρ_{av} (*blue line*), typical DOS ρ_{typ} (*black line*), and the ratio between the two ρ_{typ}/ρ_{av} (*red line*) as a function of the Fermi level. From (**a**)–(**e**), the disorder strength increases. (**a**) The band gap opens. (**b**) The band gap closes. (**c**) The mobility gap opens and the band gap disappears. (**d**) Either the band gap or the mobility gap closes. (**e**) The averaged DOS becomes flat in the strong disorder limit

both phases can be described by the Z_2 index, and belong to the same topological class. However, the disorder breaks the translational invariance. They are distinct if electrons in the bulk are localized or not.

11.7 Further Reading

- J. Li, R.L. Chu, J.K. Jain, S.Q. Shen, Topological Anderson insulator. Phys. Rev. Lett. **102**, 136806 (2009)
- H. Jiang, L. Wang, Q.-F. Sun, X.C. Xie, Numerical study of the topological Anderson insulator in HgTe/CdTe quantum wells. Phys. Rev. B **80**, 165316 (2009)
- C.W. Groth, M. Wimmer, A.R. Akhmerov, J. Tworzydło, C.W.J. Beenakker, Theory of the Topological Anderson insulator. Phys. Rev. Lett. **103**, 196805 (2009)
- H.-M. Guo, G. Rosenberg, G. Refael, M. Franz, Topological Anderson insulator in three dimensions. Phys. Rev. Lett. **105**, 216601 (2010)
- E. Prodan, Three-dimensional phase diagram of disordered HgTe/CdTe quantum spin-Hall wells. Phys. Rev. B **83**, 195119 (2011)
- D.W. Xu, J.J. Qi, J. Liu, V. Sacksteder IV, X.C. Xie, H. Jiang, Phase structure of the topological Anderson insulator. Phys. Rev. B **85**, 195140 (2012)

References

1. B. Zhou, H.Z. Lu, R.L. Chu, S.Q. Shen, Q. Niu, Phys. Rev. Lett. **101**, 246807 (2008)
2. J. Li, R.L. Chu, J.K. Jain, S.Q. Shen, Phys. Rev. Lett. **102**, 136806 (2009)
3. R. Landauer, Philos. Mag. **21**, 863 (1970)
4. M. Büttiker, IBM J. Res. Dev. **32**, 317 (1988)
5. F.D.M. Haldane, Phys. Rev. Lett. **61**, 2015 (1988)
6. E. Prodan, Phys. Rev. B **83**, 195119 (2011)
7. B.A. Bernevig, T.L. Hughes, S.C. Zhang, Science **314**, 1757 (2006)
8. C.L. Kane, E.J. Mele, Phys. Rev. Lett. **95**, 146802 (2005)
9. H.M. Guo, G. Rosenberg, G. Refael, M. Franz, Phys. Rev. Lett. **105**, 216601 (2010)
10. C.W. Groth, M. Wimmer, A.R. Akhmerov, J. Tworzydlo, C.W.J. Beenakker, Phys. Rev. Lett. **103**, 196805 (2009)
11. A. Weiße, G. Wellein, A. Alvermann, H. Fehske, Rev. Mod. Phys. **78**, 275 (2006)
12. L.W. Wang, Phys. Rev. B **49**, 10154 (1994)
13. S. Sota, M. Itoh, J. Phys. Soc. Jpn. **76**, 054004 (2007)
14. Y.Y. Zhang, R.L. Chu, F.C. Zhang, S.Q. Shen, Phys. Rev. B **85**, 035107 (2012)

Chapter 12
Summary: Symmetry and Topological Classification

Abstract For noninteracting electron systems, symmetry classification has already exhausted all possible topological insulators and superconductors: each dimension has five possible topological phases.

Keywords Topological classification • Symmetry class • Time reversal symmetry • Particle-hole symmetry

12.1 Ten Symmetry Classes for Noninteracting Fermion Systems

Following Altland and Zirnbauer [1, 2], all possible symmetry classes of random matrix, which can be interpreted as Hamiltonian of some non-interacting fermionic systems, can be systematically enumerated: there are ten symmetry classes in total. All classes are sets of Hamiltonian with specific transformation properties under some discrete symmetries.

Consider a general system of noninteracting fermions, which is described by a second quantized Hamiltonian,

$$H = \sum_{A,B} \psi_A^\dagger H_{A,B} \psi_B, \qquad (12.1)$$

where $\psi_A^\dagger$ and ψ_B are the creation and annihilation operators of fermions and satisfy the relation

$$\left\{ \psi_A^\dagger, \psi_B \right\} = \delta_{A,B}. \qquad (12.2)$$

The subscripts A and B can be collective indices. For example, for a system of electrons on a lattice, they are $A = (i, \sigma)$, which represent the electron with spin σ on the lattice site i. In this case, $H_{A,B}$ is a square matrix. The symmetries of the

S.-Q. Shen, *Topological Insulators: Dirac Equation in Condensed Matters*,
Springer Series in Solid-State Sciences 174, DOI 10.1007/978-3-642-32858-9_12,
© Springer-Verlag Berlin Heidelberg 2012

Table 12.1 Ten symmetry classes following the random matrix ensembles [3]

Symmetry classes	Discrete symmetry relation
A	$H = H^\dagger$
AI	$\epsilon_c = +1; \eta_c = +1$
AII	$\epsilon_c = +1; \eta_c = -1$
C	$\epsilon_c = -1; \eta_c = -1$
D	$\epsilon_c = -1; \eta_c = +1$
AIII	$P^2 = 1$
DI	$P^2 = 1; \epsilon_c = \pm1; \eta_c = +1; PCP^T = C$
CII	$P^2 = 1; \epsilon_c = \pm1; \eta_c = -1; PCP^T = C$
CI	$P^2 = 1; \epsilon_c = \pm1; \eta_c = \pm1; PCP^T = -C$
DIII	$P^2 = 1; \epsilon_c = \mp1; \eta_c = \pm1; PCP^T = -C$

Hamiltonian mean that the Hamiltonian H is related to $-H$, its transpose H^T, and its complex conjugation H^*, respectively. We demand that these transformations are implemented by unitary transformation and that their actions on the Hamiltonian square to one. Hence, we consider the following transformations:

P symmetry: $\quad H = -PHP^{-1}$ where $PP^\dagger = P^2 = 1$

C symmetry: $\quad H = \epsilon_c CH^TC^{-1}$ where $CC^\dagger = 1$ and $C^T = \eta_c C$ ($\epsilon_c = \pm1$ and $\eta_c = \pm1$)

K symmetry: $\quad H = \epsilon_k KH^*K^{-1}$ where $KK^\dagger = 1$ and $K^T = \eta_k K$ ($\epsilon_k = \pm1$ and $\eta_k = \pm1$)

Type P symmetry is commonly referred to as chirality symmetry, C expresses as the particle-hole symmetry, and K time reversal symmetry. For Hermitian Hamiltonians, $H = H^\dagger = (H^*)^T$. Thus, $H^T = H^*$, and C and K are identical. We shall only talk about C symmetry, where $\epsilon_c = +1$ will be interpreted as time reversal symmetry and $\epsilon_c = -1$ will be referred to as particle-hole symmetry.

An ensemble of Hamiltonian without any constraint other than being Hermitian is called the unitary symmetry class (A class). If the Hamiltonian possesses P symmetry, it is called the chiral unitary classes (AIII class). For C symmetry, we have four classes of $\epsilon_c = \pm1$ and $\eta_c = \pm1$. If the Hamiltonian possesses both P and C symmetries, then it automatically has another C-type symmetry C':

$$H = \epsilon'_c C'HC'^{-1} \tag{12.3}$$

where $C' = PC$ and $\epsilon'_c = -\epsilon_c$. Since C' can be interpreted as a time reversal symmetry if $\epsilon_c = -1$ or a particle-hole symmetry if $\epsilon_c = +1$, the classes with both P and C symmetries thus automatically have chirality, time reversal, and particle-hole symmetry. As a result, we have ten symmetry classes related to P and C symmetries as listed in Table 12.1.

Alternatively, the system can also be classified according to time reversal symmetry (TRS) Θ and the particle-hole symmetry (PHS) Υ [4]. TRS Θ can be represented by an anti-unitary operator on a Hilbert space, which is written as product of complex conjugate operator K and unitary operator C, $\Theta = KC$,

$$\Theta H \Theta^{-1} = H, \tag{12.4}$$

or the system is invariant under time reversal symmetry if and only if the complex conjugation of the Hamiltonian is equal to itself up to a unitary operator

$$\Theta : \mathbf{C}^\dagger H^*_{A,B} \mathbf{C} = +H_{A,B}. \tag{12.5}$$

Thus, for time reversal symmetry, the Hamiltonian can be (i) not time reversal invariant, in which we take $t = 0$; (ii) time reversal invariant, but the square of the time reversal operator is $+1$, $\Theta^2 = 1$, in which we take $t = 1$, for example, a spinless or integer spin system; and (iii) time reversal invariant, but the square of the time reversal operator is equal to -1, $\Theta^2 = -1$, in which we take $t = -1$. For example, a half-odd-integer spin system. So totally it has three possible cases, $t = 0, +1$, and -1.

Particle-hole symmetry (PHS) Υ can be expressed in terms of H:

$$\Upsilon H \Upsilon^{-1} = -H, \tag{12.6}$$

where $\Upsilon = KV$ or the system is invariant under time reversal symmetry if and only if the complex conjugation of the Hamiltonian is equal to a minus itself up to a unitary operator V,

$$\Upsilon : V^\dagger H^*_{A,B} V = -H_{A,B}. \tag{12.7}$$

Thus, for a particle-hole symmetry, the Hamiltonian can be (i) not particle-hole invariant, in which we take $v = 0$; (ii) particle-hole invariant, but the square of the particle-hole operator is $+1$, $\Upsilon^2 = 1$, in which we take $v = 1$; and (iii) particle-hole invariant, but the square of the particle-hole operator is equal to -1, $\Upsilon^2 = -1$, in which we take $v = -1$. So totally it has three possible cases, $v = 0, +1$, and -1.

Thus, there are at least 3×3 possible ways for a Hamiltonian to respond to time reversal and particle-hole operation. In addition, the product of TRS and PHS gives SLS = TRS × PHS, often referred to as sublattice or chiral symmetry. The assignment (TRS, PHS) = $(0, 0)$ allows SLS to be either present (SLS = 1) or absent (SLS = 0). Therefore, one obtains ten symmetry classes by combining time reversal symmetry and particle-hole symmetry together.

12.2 Physical Systems and the Symmetry Classes

12.2.1 Standard (Wigner-Dyson) Classes

Class A: The Hamiltonian which possesses neither time reversal symmetry nor the particle-hole symmetry belongs to the unitary symmetry class, that is, class A. For example, a two-dimensional electron gas in an external magnetic field.

Class AI: The Hamiltonian of integer spin or spinless particles which possesses
time reversal symmetry belongs to the orthogonal symmetry. In this case, $\Theta^2 =$
$+1$ and

$$H^T = H. \tag{12.8}$$

Class AII: The Hamiltonian of spin-$\frac{1}{2}$ particles which possesses time reversal
symmetry belongs to the symplectic symmetry. In this case, $\Theta^2 = -1$. For
example, an electron system with spin-orbit coupling

$$i\sigma_y H^T(-i\sigma_y) = H. \tag{12.9}$$

12.2.2   Chiral Classes

Symmetry classes of Hamiltonian possessing a P-type symmetry are conventionally
called chiral symmetry. In complete analog with the standard (Wigner-Dyson)
classes, there are three types of chiral symmetries:

Class AIII: The ensemble of chiral Hamiltonian without any other constraint is
called chiral unitary class.

Class CII: The ensemble of chiral Hamiltonian with time reversal symmetry and
$\Theta^2 = -1$ is called chiral symplectic class.

Class DI: The ensemble of chiral Hamiltonian with time reversal symmetry and
$\Theta^2 = +1$ is called chiral orthogonal class.

12.2.3   Bogoliubov-de Gennes (BdG) Classes

We consider a general form of a Bogoliubov-de Gennes Hamiltonian,

$$H = \frac{1}{2}(c^\dagger, c)\begin{pmatrix} \Xi & \Delta \\ -\Delta^* & -\Xi^T \end{pmatrix}\begin{pmatrix} c \\ c^\dagger \end{pmatrix} \tag{12.10}$$

where $\Xi = \Xi^\dagger$ as required by the Hermiticity of the Hamiltonian $H^\dagger = H$ and
$\Delta = -\Delta^T$ for Fermi statistics. c can be for either spinless fermions or spin-$\frac{1}{2}$
electron $c = (c_\uparrow, c_\downarrow)$.

BdG Hamiltonian can be classified into four subclasses: C and CI are primarily
relevant to spin-singlet superconductor, while D and DIII are primarily relevant to
spin-triplet superconductor.

Class D: $t_x H^T t_x = -H$ such as $p \pm ip$ wave pairing superconductor

$$H = \frac{1}{2}\sum_k (c_k^\dagger, c_{-k})\begin{pmatrix} \epsilon_k - \mu & \Delta_0(k_x \pm ik_y) \\ \Delta_0(k_x \mp ik_y) & -\epsilon_k + \mu \end{pmatrix}\begin{pmatrix} c_k \\ c_{-k}^\dagger \end{pmatrix}. \tag{12.11}$$

Class DIII: $t_x H^T t_x = -H$ and $i s_y H^T (-i s_y) = H$ such as superposition of
  $p + ip$ and $p - ip$ wave pairing superconductor
Class C: $r_y H^T r_y = -H$ such as $d \pm id$ wave pairing superconductor
Class CI: $H^* = H$ such as $d_{x^2-y^2}$ or d_{xy} wave pairing superconductor

Note that t_α, s_α, and τ_α are all the Pauli matrices.

12.3 Characterization in the Bulk

Following Schnyder et al. [4], we discuss the bulk characteristics of topological
insulator based on the spectral projection operator. In the presence of translational
invariance, the ground states of noninteracting fermion systems can be constructed
as a filled Fermi sea in the first Brillouin zone. From the eigenvalue equation in the
band theory,

$$H(k) |u_n(k)\rangle = E_n(k) |u_n(k)\rangle, \qquad (12.12)$$

the projection operator onto the filled Bloch states at a fixed k is defined as

$$P(k) = \sum_{n \in \text{filled}} |u_n(k)\rangle \langle u_n(k)|. \qquad (12.13)$$

Then it is convenient to define

$$Q(k) = 2P(k) - 1 \qquad (12.14)$$

which satisfies the relation

$$Q^2 = 1, Q^\dagger = Q \qquad (12.15)$$

and

$$\text{Tr} Q = m - n \qquad (12.16)$$

where m is the number of the filled states and n is the number of empty states.
Depending on the symmetry class, additional condition may be imposed on Q.
Without any such further conditions, the projector takes values in the so-called
Grassmannian $G_{m,m+n}(C)$: the set of eigenvectors as a unitary matrix, a member
of $U(m + n)$. Once we consider a projection onto the filled Bloch states, we have a
gauge symmetry $U(m)$. Similarly we have $U(n)$ for empty Bloch states. Thus, each
projector is described by an element of the coset

$$U(m + n)/ [U(m) \times U(n)] \simeq G_{m,m+n}(C) \simeq G_{n,m+n}(C). \qquad (12.17)$$

Since

$$Q(k) |u_n(k)\rangle = \begin{cases} + |u_n(k)\rangle & \text{if } n \text{ is filled,} \\ - |u_n(k)\rangle & \text{if } n \text{ is empty,} \end{cases} \qquad (12.18)$$

Table 12.2 Ten symmetry classes of single-particle Hamiltonian and possible topologically nontrivial ground state characterized by Z and Z_2 invariant. Z represents the group of an integer, and Z_2 represents the group of $(0, 1)$ or $(-1, +1)$ (Adapted from [4])

		TRS	PHS	SLS	$d = 1$	2	3
Standard	A (unitary)	0	0	0	–	Z	–
	AI (orthogonal)	+1	0	0	–	–	–
	AII (symplectic)	−1	0	0	–	Z_2	Z_2
Chiral	AIII (unitary)	0	0	1	Z	–	Z
	BDI (orthogonal)	+1	+1	1	Z	–	–
	CII (symplectic)	−1	−1	1	Z	–	Z_2
BdG	D	0	+1	0	Z_2	Z	–
	C	0	−1	0	–	Z	–
	DIII	−1	+1	1	Z_2	Z_2	Z
	CI	+1	−1	1	–	–	Z

an element of $G_{m,m+n}(C)$ can be written as

$$Q = U\Lambda U^\dagger, \quad \Lambda = \mathrm{diag}(1_m, -1_n), \tag{12.19}$$

and $U \in U(m + n)$. Imposing additional symmetry will prohibit certain type of maps from Brillouin zone to the space of projectors.

12.4 Five Types in Each Dimension

An element of the set of projectors within a given symmetry cannot be continuously deformed into any others without closing the energy gap between two bands, which is related to the homotopy group of the topological space of projectors. For example, the two-dimensional homotopy group is $\pi_2[G_{m,m+n}(C)] = Z$, implying that the projectors are classified by an integer or the Chern number. Possible topologically nontrivial phases with discrete symmetries are listed in Table 12.2.

12.5 Conclusion

Topological classification has exhausted all possible topological insulators and superconductors. The topological phases exist from one dimension to three dimensions and from insulators to superconductors. Some materials have been known for a long time. The topological properties of some materials were only acknowledged in recent years.

As a conclusion of this book, we can say that

each topological insulator or superconductor is governed by one modified Dirac equation.

12.6   Further Reading

- A.P. Schnyder, S. Ryu, A. Furusaki, A.W.W. Ludwig, Classification of topological insulators and superconductors in three spatial dimensions. Phys. Rev. B **78**, 195125 (2008)
- S. Ryu, A.P. Schnyder, A. Furusaki, A.W.W. Ludwig, Topological insulators and superconductors: ten-fold way and dimensionality hierarchy. New J. Phys. **12**, 065010 (2010)
- A. Kitaev, Periodic table for topological insulators and superconductors, in ed. by V. Lebedev, M. Feigel'Man. American Institute of Physics Conference Series, vol. 1134, (2009), pp. 22
- D. Bernard, A LeClair, A classification of two-dimensional random Dirac fermions. J. Phys. A Math. Gen. **35**, 2555 (2002)

References

1. A. Altland, M. Zirnbauer, Phys. Rev. B **55**, 1142 (1997)
2. M. Zirnbauer, J. Math. Phys. **37**, 4986 (1996)
3. D. Bernard, A. LeClair, J. Phys. A Math. Gen. **35**, 2555 (2002)
4. A.P. Schnyder, S. Ryu, A. Furusaki, A.W.W. Ludwig, Phys. Rev. B **78**, 195125 (2008)

Appendix A
Derivation of Two Formulae

A.1 Quantization of the Hall Conductance

In this section, we present a proof that the Hall conductance is quantized to be $\nu e^2/h$ (ν is an integer) in Eq. (4.51). For simplicity, we drop the band index first. From the definition of the Berry curvature, the Hall conductance is expressed as

$$\sigma_{xy} = \frac{e^2}{h} \frac{1}{2\pi} \int_0^{2\pi} dk_x \int_0^{2\pi} dk_y [\nabla_{\mathbf{k}} \times \mathbf{A}(k_x, k_y)]_z, \tag{A.1}$$

where the lattice constant is taken to be unit. Therefore, the conductance is determined by the Berry curvature integrated over the reduced Brillouin zone.

To evaluate the surface integral, the Stokes' theorem can be applied with the condition that the surface is simply connected. To this end, we illustrate the formation of the torus from a rectangle with the periodic boundary condition as shown in Fig. A.1. In this way, the surface integral can be reduced to a line integral around the first Brillouin zone,

$$\begin{aligned}
\sigma_{xy} &= \frac{e^2}{h} \frac{1}{2\pi} \int_0^{2\pi} dk_x \int_0^{2\pi} dk_y \left[\partial_{k_x} \mathbf{A}_y(k_x, k_y) - \partial_{k_y} \mathbf{A}_x(k_x, k_y) \right] \\
&= \frac{e^2}{h} \frac{1}{2\pi} \int_0^{2\pi} dk_y \left[\mathbf{A}_y(2\pi, k_y) - \mathbf{A}_y(0, k_y) \right] \\
&\quad - \frac{e^2}{h} \frac{1}{2\pi} \int_0^{2\pi} dk_x \left[\mathbf{A}_x(k_x, 2\pi) - \mathbf{A}_x(k_x, 0) \right].
\end{aligned} \tag{A.2}$$

Recalling that $|u(k_x, 0)\rangle$ and $|u(k_x, 2\pi)\rangle$ actually represent the same physical state due to the periodicity in the reciprocal vector space, which can only differ by a phase factor, $|u(k_x, 2\pi)\rangle = \exp[i\theta_x(k_x)]|u(k_x, 0)\rangle$, one has

S.-Q. Shen, *Topological Insulators: Dirac Equation in Condensed Matters*,
Springer Series in Solid-State Sciences 174, DOI 10.1007/978-3-642-32858-9,
© Springer-Verlag Berlin Heidelberg 2012

Fig. A.1 The equivalence of
the first Brillouin zone and a
torus: **(a)** A rectangle of the
first Brillouin zone with
periodic boundary conditions
(b) the rectangle is rolled into
a tube along the k_y direction.
(c) The tube is rolled into a
torus along the k_x direction.
The four corners of the
rectangle are actually the one
point in the torus surface

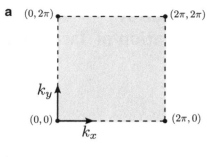

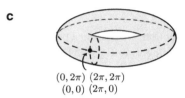

$$\mathbf{A}_x(k_x, 2\pi) = \langle u(k_x, 2\pi) | i\, \partial_{k_x} | u(k_x, 2\pi) \rangle$$

$$= -\partial_{k_x} \theta_x(k_x) + \mathbf{A}_x(k_x, 0). \tag{A.3}$$

Similarly, taking $|u(2\pi, k_y)\rangle = \exp[i\,\theta_y(k_y)] |u(0, k_y)\rangle$, one obtains

$$\mathbf{A}_y(2\pi, k_y) = -\partial_{k_y} \theta_y(k_y) + \mathbf{A}_y(0, k_y). \tag{A.4}$$

$\theta_x(k_x)$ and $\theta_y(k_y)$ are smooth functions. Using these two relations, the integral is
reduced to

$$\sigma_{xy} = \frac{e^2}{h} \frac{1}{2\pi} \int_0^{2\pi} dk_y \left[-\partial_{k_y} \theta_y(k_y) \right] + \frac{e^2}{h} \frac{1}{2\pi} \int_0^{2\pi} dk_x \left[\partial_{k_x} \theta_x(k_x) \right]$$

$$= \frac{e^2}{h} \frac{1}{2\pi} \left[\theta_y(0) - \theta_y(2\pi) + \theta_x(2\pi) - \theta_x(0) \right]. \tag{A.5}$$

On the torus surface of the first Brillouin zone, the four wave states $|u(0,0)\rangle$,
$|u(0, 2\pi)\rangle$, $|u(2\pi, 0)\rangle$, and $|u(2\pi, 2\pi)\rangle$ actually represent the same states (see in
Fig. A.1). Using the phase matching relations of these states,

$$e^{i\theta_x(0)} |u(0, 2\pi)\rangle = |u(0, 0)\rangle, \tag{A.6a}$$

$$e^{i\theta_x(2\pi)} |u(2\pi, 2\pi)\rangle = |u(2\pi, 0)\rangle, \tag{A.6b}$$

$$e^{i\theta_y(0)}|u(2\pi, 0)\rangle = |u(0, 0)\rangle, \tag{A.6c}$$

$$e^{i\theta_y(2\pi)}|u(2\pi, 2\pi)\rangle = |u(0, 2\pi)\rangle, \tag{A.6d}$$

one obtains

$$|u(0, 0)\rangle = e^{i[\theta_x(0) + \theta_y(2\pi) - \theta_x(2\pi) - \theta_y(0)]}|u(0, 0)\rangle. \tag{A.7}$$

The single valuedness of $|u(0, 0)\rangle$ requires that the exponent must be an integer multiple of 2π, that is,

$$\theta_x(0) + \theta_y(2\pi) - \theta_x(2\pi) - \theta_y(0) = 2\nu\pi \tag{A.8}$$

with an integer ν (including 0). Therefore, the Hall conductance must be quantized when the band is fully filled. This integer ν is called Thouless-Kohmoto-Nightingale-Nijs (TKNN) number or the first Chern number, which characterizes the topological structure of the Bloch states $|u(k_x, k_y)\rangle$ in the parameter space (k_x, k_y).

A.2   A Simple Formula for the Hall Conductance

A simple two-band model has a general form in terms of the Pauli matrices σ_α,

$$H(\mathbf{k}) = \epsilon(\mathbf{k}) + \sum_{\alpha=1,2,3} d_\alpha(\mathbf{k})\sigma_\alpha. \tag{A.9}$$

The energy spectra of the model are

$$E_\pm(\mathbf{k}) = \epsilon(\mathbf{k}) \pm d(\mathbf{k}) \tag{A.10}$$

with $d(\mathbf{k}) = \sqrt{\sum_{\alpha=1,2,3}|d_\alpha(\mathbf{k})|^2}$, and the corresponding eigenstates are

$$|\mathbf{k}, +\rangle = \begin{pmatrix} \cos\frac{\theta}{2}e^{-i\phi} \\ \sin\frac{\theta}{2} \end{pmatrix}, \tag{A.11a}$$

$$|\mathbf{k}, -\rangle = \begin{pmatrix} \sin\frac{\theta}{2}e^{-i\phi} \\ -\cos\frac{\theta}{2} \end{pmatrix}, \tag{A.11b}$$

where $\theta = \arccos\frac{d_z(\mathbf{k})}{d(\mathbf{k})}$ and $\phi = \arctan\frac{d_y(\mathbf{k})}{d_x(\mathbf{k})}$.

In electric conduction, the conductivity $\sigma_{\alpha\beta}$ is defined as

$$J_\alpha(\mathbf{r}, t) = \sum_\beta \sigma_{\alpha\beta}(\mathbf{q}, \omega) \Xi_\beta \exp[i(\mathbf{q} \cdot \mathbf{r} - \omega t)], \tag{A.12}$$

where $J_\alpha(\mathbf{r}, t)$ is the electric current and $\Xi_\beta \exp[i(\mathbf{q} \cdot \mathbf{r} - \omega t)]$ is the electric field. In the linear response theory, the Kubo formula for the Hall conductance gives

$$\sigma_{xy}(\mathbf{q}, \omega) = +\frac{i}{\omega}\Pi_{xy}(\mathbf{q}, \omega) \tag{A.13}$$

with the retarded correlation function of the current operator $J_x(\mathbf{q}, t)$ and $J_y(\mathbf{q}, t')$

$$\Pi_{xy}(\mathbf{q}, \omega) = -\frac{i}{V}\int_{-\infty}^{+\infty} dt\theta(t - t')e^{i\omega(t-t')} \langle\psi|[J_x(\mathbf{q}, t), J_y(\mathbf{q}, t')]|\psi\rangle, \tag{A.14}$$

where V is the volume of the system. The dc conductivity is obtained by taking the limit $\mathbf{q} \to \mathbf{0}$ and then $\omega \to 0$,

$$\sigma_{xy} = \lim_{\omega \to 0}\lim_{q \to 0} \sigma_{xy}(\mathbf{q}, \omega). \tag{A.15}$$

Usually the retarded correlation function can be calculated in the Matsubara formalism

$$\Pi_{xy}^M(i\omega_\nu) = \frac{1}{V}\frac{1}{\beta}\sum_{k,\nu'}\text{Tr}\left\{J_x(\mathbf{k})G[\mathbf{k}, i(\omega_\nu + \omega_{\nu'})]J_y(k)G[\mathbf{k}, i\omega_{\nu'}]\right\} \tag{A.16}$$

with frequencies $\omega_\nu = 2\nu\pi/\beta$ and $\omega_{\nu'} = (2\nu' + 1)\pi/\beta$ ($\beta = k_BT$). The Matstubara-Green's function is given by

$$G(\mathbf{k}, i\omega_\nu) = [i\omega_\nu - H(\mathbf{k})]^{-1}$$

$$\equiv \frac{P_+}{i\omega_\nu - E_+(\mathbf{k})} + \frac{P_-}{i\omega_\nu - E_-(\mathbf{k})} \tag{A.17}$$

with

$$P_\pm = \frac{1}{2}\left[1 \pm \sum_{\alpha=1,2,3}\frac{d_\alpha(\mathbf{k})\sigma_\alpha}{d}\right]. \tag{A.18}$$

Using the frequency summation over $i\omega_{\nu'}$,

$$\frac{1}{\beta}\sum_{\nu'}\frac{1}{i(\omega_\nu + \omega_{\nu'}) - E_n}\frac{1}{i\omega_{\nu'} - E_m} = \frac{f_{\mathbf{k},m} - f_{\mathbf{k},n}}{i\omega_\nu + E_m(\mathbf{k}) - E_n(\mathbf{k})}, \tag{A.19}$$

where the Dirac-Fermi distribution function $f_{\mathbf{k},n} = 1/\{1 + \exp[\beta(E_n(\mathbf{k}) - \mu)]\}$, one obtains

$$\Pi_{xy}^{M}(\omega_v) = \frac{1}{V} \sum_{\mathbf{k},n,n'} \langle \mathbf{k},n| J_x(\mathbf{k}) |\mathbf{k},n'\rangle \langle \mathbf{k},n'| J_y(\mathbf{k}) |\mathbf{k},n\rangle \frac{f_{\mathbf{k},n} - f_{\mathbf{k},n'}}{i\omega_v + E_n(\mathbf{k}) - E_{n'}(\mathbf{k})}.$$
(A.20)

Its analytical continuation to the retarded function is realized by replacing $i\omega_n \to \hbar\omega + i\epsilon$,

$$\Pi_{xy}^{M}(\omega_v) \to \Pi_{xy}^{R}(\omega).$$
(A.21)

Using l'Hôspital's rule,

$$\lim_{\omega \to 0} \frac{\operatorname{Im}(\Pi_{xy}^{R}(\omega))}{\omega} = \operatorname{Im}\left(\frac{d\Pi_{xy}^{R}(\omega)}{d\omega}\right)_{\omega=0},$$
(A.22)

and

$$\lim_{\omega \to 0} \frac{d}{\hbar d\omega}\left[\frac{1}{\hbar\omega + i\epsilon + E_n - E_{n'}}\right] = -\frac{1}{(E_n - E_{n'})(E_n - E_{n'} + i\epsilon)},$$
(A.23)

the Kubo formula for the dc Hall conductivity can be written as

$$\sigma_{xy} = \frac{\hbar}{V} \lim_{\epsilon \to 0^+} \sum_{k,n \neq n'} \frac{(f_{\mathbf{k},n} - f_{\mathbf{k},n'})\operatorname{Im}\left(\langle \mathbf{k},n| J_x(\mathbf{k}) |\mathbf{k},n'\rangle \langle \mathbf{k},n'| J_y(\mathbf{k}) |\mathbf{k},n\rangle\right)}{(E_n(\mathbf{k}) - E_{n'}(\mathbf{k}))(E_n(\mathbf{k}) - E_{n'}(\mathbf{k}) + i\epsilon)}.$$
(A.24)

From the model in Eq. (A.9), the current operator $J_i(\mathbf{k}) = -ev_i(\mathbf{k})$ is given by

$$J_i(\mathbf{k}) = -\frac{e}{\hbar}\partial_{k_i} H(\mathbf{k}) = -\frac{e}{\hbar}\left(\partial_{k_i}\epsilon(\mathbf{k}) + \sum_{\alpha=1,2,3}\partial_{k_i}d_\alpha(\mathbf{k})\sigma_\alpha\right).$$
(A.25)

For $n \neq n'$, one has

$$\langle \mathbf{k},n| J_i(\mathbf{k}) |\mathbf{k},n'\rangle = -\frac{e}{\hbar}\sum_{\alpha=1,2,3}\partial_{k_i}d_\alpha(\mathbf{k}) \langle \mathbf{k},n| \sigma_\alpha |\mathbf{k},n'\rangle.$$
(A.26)

Furthermore,

$$\operatorname{Im}\left(\langle \mathbf{k},n| \sigma_\alpha |\mathbf{k},-n\rangle \langle \mathbf{k},-n| \sigma_\beta |\mathbf{k},n\rangle\right) = n\epsilon_{\alpha\beta\gamma}\frac{d_\gamma(\mathbf{k})}{d(\mathbf{k})}.$$
(A.27)

We limit our discussion in the case that two levels do not cross in the whole momentum space such that $\epsilon \to 0^+$ can be taken before the integral of $\mathbf{k}$. Thus, the conductance can be expressed as

$$\sigma_{xy} = \frac{1}{2\Omega}\frac{e^2}{\hbar}\sum_k \epsilon_{\alpha\beta\gamma}\frac{[\partial_{k_x}d_\alpha(\mathbf{k})]\left[\partial_{k_y}d_\beta(\mathbf{k})\right]d_\gamma(\mathbf{k})}{d^3(\mathbf{k})}(f_{\mathbf{k},+} - f_{\mathbf{k},-}). \qquad (A.28)$$

If there exists an energy gap between the upper and lower bands, and the lower band is fully filled, that is, $E_{\mathbf{k},-} < \mu < E_{\mathbf{k},+}$, then $f_{\mathbf{k},+} = 0$ and $f_{\mathbf{k},-} = 1$ at zero temperature. The Hall conductance has the form

$$\sigma_{xy} = -\frac{e^2}{h}\frac{1}{4\pi}\int dk_x dk_y \frac{\left(\partial_{k_x}\mathbf{d}(\mathbf{k})\times\partial_{k_y}\mathbf{d}(\mathbf{k})\right)\cdot\mathbf{d}(\mathbf{k})}{d^3(\mathbf{k})}. \qquad (A.29)$$

Appendix B
Time Reversal Symmetry

Time reversal symmetry is the invariance of physical laws under time reversal transformation. The terminology was first introduced by E. Wigner in 1932.

B.1 Classical Cases

Let us first look at the classical case: a motion of particle subjected to a certain force. Its trajectory is given by the Newtonian equation of motion,

$$m\frac{d^2\mathbf{r}}{dt^2} = -\nabla V(r). \tag{B.1}$$

If $\mathbf{r}(t)$ is the solution of the equation, then $\mathbf{r}(-t)$ is also the solution of the equation. In another words, when we make a transformation $t \to -t$, the Newtonian equation of motion keeps unchanged. Of course we should notice the change of the boundary condition or initial conditions for the problem.

Maxwell's equations and the Lorentz force $\mathbf{F} = -e(\mathbf{E} + \mathbf{v} \times \mathbf{B})$ are invariant under the time reversal provided that

$$\mathbf{v} \to -\mathbf{v}, \mathbf{j} \to -\mathbf{j}, \rho \to \rho, \tag{B.2}$$

$$\mathbf{B} \to -\mathbf{B}, \mathbf{E} \to \mathbf{E}. \tag{B.3}$$

Maxwell's equations are

$$\nabla \cdot \mathbf{D} = \rho, \nabla \times \mathbf{E} + \frac{\partial \mathbf{B}}{\partial t} = 0, \tag{B.4a}$$

$$\nabla \cdot \mathbf{B} = 0, \nabla \times \mathbf{H} - \frac{\partial \mathbf{D}}{\partial t} = I, \tag{B.4b}$$

S.-Q. Shen, *Topological Insulators: Dirac Equation in Condensed Matters*,
Springer Series in Solid-State Sciences 174, DOI 10.1007/978-3-642-32858-9,
© Springer-Verlag Berlin Heidelberg 2012

where $\mathbf{D} = \epsilon_0 \mathbf{E} + \mathbf{P}$ and $\mathbf{H} = \mathbf{B}/\mu_0 - \mathbf{M}$. Therefore, a magnetic field changes a sign, and an electric field remains unchanged under time reversal.

In quantum mechanics, the Schrödinger equation is written as

$$i\hbar \frac{\partial \Psi(\mathbf{x}, t)}{\partial t} = \left(-\frac{\hbar^2}{2m} \nabla^2 + V \right) \Psi(\mathbf{x}, t), \tag{B.5}$$

in which the Hamiltonian at the right-hand side is invariant under the time reversal. If $\Psi(\mathbf{x}, t)$ is a solution of the equation, $\Psi(\mathbf{x}, -t)$ is not a solution of the equation because of the first-order time derivative and the imaginary sign at the left-hand side. However, $\Psi^*(\mathbf{x}, -t)$ is a solution. One can check it by using the solution of a free particle, $\Psi(\mathbf{x}, t) = c e^{i(\mathbf{p}\cdot\mathbf{x} - Et)/\hbar}$. The $\Psi(\mathbf{x}, -t) = c e^{i(\mathbf{p}\cdot\mathbf{x} + Et)/\hbar}$ is also a solution of the Schrödinger equation. However, the momentum is still $\mathbf{p}$, *not* $-\mathbf{p}$.

Definition. The transformation θ

$$|\alpha\rangle \rightarrow |\tilde{\alpha}\rangle = \theta |\alpha\rangle, |\beta\rangle \rightarrow \left|\tilde{\beta}\right\rangle = \theta |\beta\rangle \tag{B.6}$$

is said to be anti-unitary if

$$\left\langle \tilde{\beta} | \tilde{\alpha} \right\rangle = \langle \beta | \alpha \rangle^* ; \tag{B.7a}$$

$$\theta \left(c_1 |\alpha\rangle + c_2 |\beta\rangle \right) = c_1^* \theta |\alpha\rangle + c_2^* \theta |\beta\rangle . \tag{B.7b}$$

In this case, the operator θ is an anti-unitary operator. Usually an anti-unitary operator can be written as

$$\theta = UK, \tag{B.8}$$

where U is a unitary operator and K is the complex conjugation operator, which is defined as

$$K\varphi = \varphi^* K. \tag{B.9}$$

Here φ can be either a function or an operator.

B.2 Time Reversal Operator Θ

Let us denote the time reversal operator by Θ. Consider

$$|\alpha\rangle \rightarrow \Theta |\alpha\rangle, \tag{B.10}$$

where $\Theta |\alpha\rangle$ is the time-reversed state. More appropriately, $\Theta |\alpha\rangle$ should be called the motion-reversed state. For a momentum eigenstate $|\mathbf{p}\rangle$, $\Theta |\mathbf{p}\rangle$ should be $|-\mathbf{p}\rangle$ up to a possible phase factor. Θ is an anti-unitary operator. We can see this property from the Schrödinger equation of a time reversal invariant system,

$$i\hbar\frac{\partial}{\partial t}\Psi(\mathbf{x}, t) = H\Psi(\mathbf{x}, t),\tag{B.11}$$

provided that $\Theta i\,\Theta^{-1} = -i$ and $\Theta\frac{\partial}{\partial t}\Theta^{-1} = \frac{\partial}{\partial(-t)}$. The transformed momentum operator $\mathbf{p}$, the position $\mathbf{x}$, and the angular momentum $\mathbf{J}$ are

$$\Theta\mathbf{p}\Theta^{-1} = -\mathbf{p},\tag{B.12a}$$

$$\Theta\mathbf{x}\Theta^{-1} = \mathbf{x},\tag{B.12b}$$

$$\Theta\mathbf{J}\Theta^{-1} = -\mathbf{J}\ (= \mathbf{x} \times \mathbf{p}).\tag{B.12c}$$

Note that for $\mathbf{p} = -i\hbar\frac{d}{dx}$, $\Theta\mathbf{p}\Theta^{-1} = -\mathbf{p}$.

From the spherical harmonic $Y_l^m(\theta, \phi)$, one has

$$Y_l^m(\theta, \phi) \rightarrow \left(Y_l^m(\theta, \phi)\right)^* = (-1)^m Y_l^{-m}(\theta, \phi).\tag{B.13}$$

Therefore, the eigenstate $|l, m\rangle$ of the orbital angular momentum and its z-component has the relation

$$\Theta\,|l, m\rangle = (-1)^m\,|l, -m\rangle\,.\tag{B.14}$$

B.3   Time Reversal for a Spin-$\frac{1}{2}$ System

Under the time reversal, $t \rightarrow -t$. Applying the time reversal operation twice, can we go back to the original states? Yes, but Θ^2 is not always equal to 1. For a spin-$\frac{1}{2}$ system,

$$\Theta\sigma_\alpha\Theta^{-1} = -\sigma_\alpha,\tag{B.15}$$

where $\alpha = x, y, z$. Note that

$$\sigma_y\sigma_x\sigma_y = -\sigma_x,\tag{B.16a}$$

$$\sigma_y\sigma_y\sigma_y = +\sigma_y,\tag{B.16b}$$

$$\sigma_y\sigma_z\sigma_y = -\sigma_z.\tag{B.16c}$$

By convention, σ_y is taken to be purely imaginary as in Eq. (2.4), and σ_x and σ_z are real. We have $K\sigma_y = -\sigma_y K$ and $K\sigma_{x,z} = \sigma_{x,z}K$. Therefore, the time reversal operator can be constructed by combining σ_y and the complex conjugation operator K,

$$\Theta = i\sigma_y K.\tag{B.17}$$

Its inverse matrix is

$$\Theta^{-1} = -\Theta = -i\sigma_y K. \tag{B.18}$$

One can check the relation

$$\Theta^2 = -1. \tag{B.19}$$

Consider the eigenstate $|n, +\rangle$ of $\mathbf{S} \cdot \mathbf{n}$ with the eigenvalue $+\hbar/2$,

$$|n, +\rangle = e^{-iS_z\alpha/\hbar} e^{-iS_y\beta/\hbar} |+\rangle, \tag{B.20a}$$

$$\Theta |n, +\rangle = \Theta e^{-iS_z\alpha/\hbar} e^{-iS_y\beta/\hbar} \Theta^{-1}\Theta |+\rangle. \tag{B.20b}$$

Because $\Theta S_\alpha \Theta^{-1} = -S_\alpha$ and $\Theta i \Theta^{-1} = -i$,

$$\Theta |n, +\rangle = e^{-iS_z\alpha/\hbar} e^{-iS_y\beta/\hbar} \Theta |+\rangle = e^{-iS_z\alpha/\hbar} e^{-iS_y\beta/\hbar} |-\rangle = |n, -\rangle, \tag{B.21}$$

where $\Theta |+\rangle = |-\rangle$ with an eigenvalue $-\frac{1}{2}$. On the other hand,

$$|n, -\rangle = e^{-iS_z\alpha/\hbar} e^{-iS_y(\pi+\beta)/\hbar} |+\rangle = e^{-iS_z\alpha/\hbar} e^{-iS_y\beta/\hbar} e^{-iS_y\pi/\hbar} |+\rangle. \tag{B.22}$$

Noting that K acting on $|+\rangle$ gives $|+\rangle$. We have

$$\Theta = e^{-i\pi S_y/\hbar} K = i\sigma_y K. \tag{B.23}$$

In general, for a system with an angular momentum operator of the eigenvalue j, the time reversal operator is

$$\Theta = i e^{-i\pi J_y} K, \tag{B.24}$$

where J_y is the y-component of orbital angular momentum operator. The operator satisfies the relation

$$\Theta^2 = (-1)^{2j}. \tag{B.25}$$

Kramers Degeneracy: The energy states for odd number of electrons in a time reversal invariant system has at least double degeneracy.

This theorem is determined by the fact that the total spin of odd number of electrons is always half of odd number of $\hbar$. The time reversal operator has always the relation $\Theta^2 = -1$.

Index

S.-Q. Shen, *Topological Insulators: Dirac Equation in Condensed Matters*,
Springer Series in Solid-State Sciences 174, DOI 10.1007/978-3-642-32858-9,
© Springer-Verlag Berlin Heidelberg 2012

CPSIA information can be obtained at www.ICGtesting.com
Printed in the USA
LVOW01*1537231114

415191LV00001B/138/P